Building a Winning Career in Engineering

Building a Winning Career in Engineering

20 Strategies for Success After College

David L. Goetsch

Upper Saddle River, New Jersey
Columbus, Ohio

Library of Congress Cataloging in Publication Data

Goetsch, David L.
 Building a winning career in engineering : 20 strategies for success
   after college / by David L. Goetsch.—1st ed.
   p. cm.
   ISBN 0-13-119211-6
 1. Engineering—Vocational guidance. 2. Career development. I. Title.
TA157.G62 2007
620.0023—dc22

                                        2006022113

Editor: Jill Jones-Renger
Editorial Assistant: Yvette Schlarman
Production Editor: Kevin Happell
Design Coordinator: Diane Ernsberger
Cover Designer: Aaron Dixon
Cover art: Aaron Dixon
Production Manager: Deidra Schwartz
Director of Marketing: David Gesell
Senior Marketing Manager: Jimmy Stephens
Senior Marketing Coordinator: Elizabeth Farrell
Marketing Coordinator: Alicia Dysert

This book was set in Goudy Sans Medium by Integra Software Service, Pvt. Ltd. It was printed and bound by
Courier Stoughton, Inc. The cover was printed by Courier Stoughton, Inc.

Pearson Education Ltd.
Pearson Education Singapore Pte. Ltd.
Pearson Education Canada, Ltd.
Pearson Education—Japan

Pearson Education Australia Pty. Limited
Pearson Education North Asia Ltd.
Pearson Educación de Mexico, S.A. de C.V.
Pearson Education Malaysia Pte. Ltd.

10 9 8 7 6 5 4 3 2 1
ISBN 0-13-119211-6

CONTENTS

CHAPTER 3
Become a Leader in Your Organization   23

CHAPTER 4
Set High Expectations for Yourself and Develop a Plan for Achieving Them   34

CHAPTER 5
Identify Positive Role Models and Learn From Them   42

CHAPTER 13
Develop and Apply a Can-Do Attitude and Seek Responsibility   124

CHAPTER 14
Don't Just Plan—Execute   131

CHAPTER 15
Become an Effective Public Speaker   138

CHAPTER 20
Persevere Through Difficult Times—Don't Quit and Never Give Up   203

APPENDICES   213

INDEX   215

INTRODUCTION

Let me begin with a disclaimer. This is NOT a book about how to find a job after finishing college. There are already plenty of those, and many of them are quite good. You can also get help finding a job from your college's career center and various Internet sites. However, the Appendices to this book contain tips that might be helpful when preparing for a job search. Rather than focus on the job search, this book focuses on how to succeed in your career after you have found a job in engineering or an engineering related profession.

Consider the following scenario. Two people of the same age go to the same college, select the same engineering major, make the same grades, and graduate with the same degree at the same time. One of them goes on to have a rewarding, successful career while the other suffers through a humdrum, mediocre career. This scenario is real. I graduated from college with the two students in question. During a career that has spanned forty years to date, I have witnessed this scenario repeated over and over again.

Many years ago, I began to wonder why people who appear so similar in background and qualifications can be so different when it comes to the success they enjoy in their careers. Are people who succeed in their careers different than those who don't? Do successful people do things differently, or is success just the product of luck, timing, or other factors beyond their control? After pondering such questions for years, I finally decided to find some answers.

THE STUDY THAT LED TO THIS BOOK

My study concerning why and how certain college graduates succeed in their careers when others don't took five years to complete. During this time, I interviewed more than two hundred highly successful college graduates who had majored in engineering and engineering-related fields. As a result of these interviews, I learned that succeeding in a career is just part—although a major part—of succeeding more broadly in life. I also learned that there are specific attitudes, characteristics, traits, and behaviors that separate successful people from mediocre and unsuccessful people. I call these attitudes, characteristics, traits, and behaviors *success strategies*.

Those interviewed during my study had strong opinions concerning why some people succeed when others don't, and they provided many interesting and insightful examples to support their opinions. The twenty success strategies recommended most frequently by study subjects became the basis for this book. In the opinions of the study subjects, these twenty strategies are at the very heart of the success they have enjoyed in their careers; they are strategies that can mean the difference between success and failure for others. Those successful professionals interviewed have applied all of the strategies recommended in this book in building winning careers for themselves, and they still apply them as a way to ensure continued success.

TWENTY SUCCESS STRATEGIES

The twenty success strategies explained in this book were recommended by all two hundred of the study subjects. Not all individuals interviewed ranked them in the same order of priority and, of course, there were other strategies recommended in addition to the twenty contained herein. However, the twenty success strategies chosen for this book are those that received by far the largest number of recommendations from study subjects. These strategies are as follows:

1. Make integrity your hallmark.
2. Become an effective communicator and listener.
3. Become a leader in your field.
4. Set high expectations for yourself and develop a plan for achieving them.
5. Identify positive role models and learn from them.
6. Be an effective team builder.
7. Be a positive change agent.
8. Project a winning image.
9. Become an effective negotiator.
10. Develop and apply self-discipline.
11. Never just pass the time—use it.
12. Become a customer-focused professional.
13. Develop and apply a positive can-do attitude.
14. Don't just plan—execute.
15. Become an effective public speaker.
16. Become a creative problem solver and critical thinker.
17. Learn to manage conflict and deal with difficult people.
18. Help others succeed—be a mentor.
19. Learn to balance work and the rest of your life.
20. Persevere through difficult times—never give up—don't quit.

PRACTICAL "HOW-TO" APPROACH

This book takes a very practical, how-to approach in explaining not just what you need to do in order to succeed, but how to actually go about it. For example, it is one thing to say that you need to become a good communicator in order to succeed in your career, but it is quite another to explain how to do that. Knowing what to do is an important step in the right direction, but knowing how to do it is an even more critical step.

THE FOUNDATION: MASTER YOUR COLLEGE SUBJECTS FIRST

This book assumes that you have already completed or will complete the critical foundational step of mastering your college coursework. The courses you take in college, especially those in your major, are the foundation upon which you will build a winning career in your profession by applying the strategies set forth in this book. Unless you first master your college subjects, the twenty strategies in this book will be applied to no avail.

The approach taken herein is that the abilities needed to succeed in a given career are not innate. In other words, you are not born with them. Rather, these success skills are just like the material in your college courses in that with work they can be learned. Each success strategy identified in the study that led to this book is the subject of a chapter in the book. What you need to do in terms of each strategy is presented at the beginning of each chapter. This presentation is then followed by an in-depth explanation of how to implement the strategy in question. As a result, this book can serve as your personal how-to manual for succeeding after college.

CASES, SUCCESS PROFILES, AND DISCUSSION QUESTIONS

Each chapter contains one or more cases that relate the stories of selected individuals. These cases illustrate the importance of the success strategy in question as well as how to apply that strategy effectively. The cases are real, but the names of the individuals, organizations, and certain circumstances in them have been changed to protect confidentiality. Most of the cases relate the stories of individuals I came to know during my career. Other cases were related to me by those interviewed during the study that led to this book.

Five success profiles have been included to illustrate how five different engineering professionals have applied the strategies recommended in this book to climb the career ladder and make it to the top in their professions. There is one such profile in Chapters 1, 3, 5, 10, 15, 16, and 20. All of the names and circumstances in the success profiles are those of the actual individuals profiled. Nothing about their life stories has been changed. I recommend that you review these profiles carefully, making special note of how these successful individuals applied the strategies recommended in this book to their lives and work. As did the individuals profiled in this book, you too can build a winning career using the twenty strategies explained herein. My hope is that in future editions of this book, I will be able to relate *your* success profile.

CHAPTER ONE

Make Integrity Your Hallmark

Integrity is consistency between actions and values . . . if individuals don't abide by their word, fundamental relationships of trust are in jeopardy.[1]

   A. Gostick

In the long run, few things will contribute more to a successful career than integrity. Notice that I said "in the long run." Several factors can make living a life of integrity difficult. One of the most pervasive of these factors is the natural human desire for immediate tangible benefits. More often than not, the tangible benefits of integrity are realized more in the long run; they are rarely immediate. The internal rewards of a life of integrity are immediate, but the tangible rewards such as recognition, promotions, salary increases, and bonuses often are not. When interviewing participants as part of the study that led to this book, I always concluded with the following question: *In your opinion, what is the ONE characteristic that is the most important factor in building a winning career?* The overwhelming majority of study participants gave the same response— integrity.

WHAT IS INTEGRITY?

Integrity is strict and consistent adherence to a set of core values and beliefs. A person with integrity adopts a positive value system, knows what beliefs make up that system, and lives according to those beliefs. Living a life of integrity means maintaining consistency between what you believe and how you live. Your integrity or lack of it will manifest itself in the choices you make on a daily basis.

WHY DON'T MORE PEOPLE LIVE A LIFE OF INTEGRITY?

All of the factors that, together, make living a life of integrity a daily challenge fall under the broad umbrella of *perceived personal interest*. I say "perceived" because life proves over and over again that in the long run the individual's personal interests are served best by integrity. However, an unfortunate fact of life is that it will often appear, at least in the short run, that your personal interests are best served by unethical choices. The personal-interest factors that most often drive unethical choices are *greed, impatience, ego, fear, expedience,* and *ambition.*

Driven by greed, a person might inflate his monthly expense report or claim to have worked hours that he didn't so as to increase his income. Driven by impatience, people might exaggerate their credentials in order to climb the career ladder faster. Driven by ego, people might sabotage the career of colleagues by circulating negative rumors about them. Driven by fear of retribution, people might turn a blind eye to the unethical behavior of a superior or a colleague. Driven by expedience people might cut corners on a project to get it done on schedule without putting in the time and effort required to do it right. Driven by misguided ambition, people might lie about their experience during an interview for a job. Living a life of integrity means constantly battling against the temptations of misguided self-interest, and winning the battle.

CHEATING AS A WAY OF LIFE?

Has cheating become an accepted fact of life in our society? Is cheating something everyone does? Have we become a society that not only looks the other way when people cheat, but actually condones or even encourages cheating? According to David Callahan, America has developed a "cheating culture."[2] Callahan cites numerous examples that support his contention that cheating in America has become an institutionalized norm rather than a rare exception.

Callahan cites such examples of endemic cheating as these: (1) psychiatrists are pressured by wealthy parents to falsely diagnose their children as learning disabled so the children will be allowed more time when taking their SAT tests for college admission (the psychiatrists give in to the pressure); (2) a researcher on the faculty of a prestigious university is hired as a consultant by a major pharmaceutical company and pressured to lie to physicians about the uses and benefits of a new drug (the researcher gives in to the pressure and lies to physicians); (3) a corporate Chief Executive Officer (CEO) is caught falsely claiming to have an MBA degree; (4) a woman keeps the receipt for the entire cost of a meal she and several other colleagues divided up so she can claim the entire cost on her expense account rather than just the part of the meal she paid for (her colleagues know what she plans to do, but no one objects); (5) thousands of people use technology to illegally download music from the Internet, thereby robbing the artists who created the music of revenue that is rightly theirs; (6) a newspaper reporter fabricates quotes and makes up stories and then turns his dishonesty into cash by writing a tell-all book damaging to his newspaper's reputation; and (7) corporate executives admit in a survey that they frequently cheat in business-related golf matches (eighty-two percent admit to cheating).[3]

Callahan's concern is not just that people cheat, but that cheating seems to be so widely accepted. He gives the following reasons for society's widespread acceptance of cheating:[4]

■ *New pressures.*   With the advent of globalization, the daily pressure to perform in today's intensely competitive workplace has become so overwhelming that our sense of right and wrong has gotten lost in a fog of desperation.

- *Winning pays.*   The size of the payoff for winning has increased in many cases to the point that some people think the end justifies the means. The end might be financial reward, a promotion, admission to a prestigious graduate program, or some other desirable reward. The means might involve cheating, lying, stealing, or other unethical methods.
- *Temptation.*   The temptation to cheat has increased in proportion to the decrease in the ability or willingness of the government and the legal system to effectively enforce regulations and laws.
- *Trickle-down corruption.*   Let's say you are a college student who, during your studies, learned about corporate executives or elected officials who won major victories by cheating. You might think, "If these executives do it, why shouldn't I?" The unethical behavior of seemingly successful people can have a trickle-down effect on others who would like to improve their lot in life.

EVENTUALLY THE TRUTH WILL COME OUT

The following example illustrates a hard truth about integrity and the lack of it. When people behave in an unethical way in order to serve their self-interests, eventually the truth will come out, and when it does the negative consequences invariably outweigh the earlier benefits. John and Joe grew up together, graduated from high school in the same class, attended the same university, selected the same major, made the same grades, and graduated from college on the same day. On the surface, John and Joe looked like carbon copies of each other. Their resumes were interchangeable. However, as it turned out, John and Joe were two very different people.

After graduating from college, John and Joe went to work at different companies in the same town. They began their careers at the same level, but within just a year it became clear that John was outpacing his friend as they both attempted to climb the career ladder. John was being promoted faster, and based on the material evidence—house, car, entertainment center, etc.—making a lot more money than Joe. At first Joe attributed John's evident success to superior performance on the job. He assumed that John was working harder and smarter than his contemporaries. He had seen John do this in high school and college. Then, one day while the two friends were having lunch, John brought up the subject of Joe's comparatively slow career advancement.

The lunch discussion turned out to be a real eye opener for Joe. He learned that John was more talented at self-promotion, office politics, backstabbing, cutting corners, and shading the truth than he was at actually doing his job. It was these things and not his talent and hard work that were catapulting John over his contemporaries on the career ladder. When Joe questioned the ethics of his behavior, John just laughed and told Joe he was naïve. John's defense was that "Everybody does it, so why shouldn't I?"

After this disturbing but revealing discussion, the two friends drifted apart. As Joe continued to slowly but surely advance in his career, he kept tabs on his old friend's meteoric progress. By the time Joe was a supervisor, John had achieved his ultimate career goal. He had made it to the top of his company as president and CEO. John became a major contributor to their alma mater and soon had the ear of local politicians. The gap between their circumstances became so pronounced that Joe began to doubt the veracity of his value system. Perhaps John was right. Maybe *integrity* was just a philosophical concept well suited to academic discussions in a college classroom, but out of place in the rough-and-tumble world of work.

As John's material success, awards, and recognition continued to pile up, Joe began to wonder if John hadn't been right all along. Maybe he *was* naïve. While Joe was struggling with this crisis of conscience, he opened his newspaper one morning and found himself looking at a photograph of John being escorted out of his company's facility by police officers. The story below the photograph summarized a long list of criminal charges alleged against John that included fraud, income tax evasion, bribery, and money laundering. When all was said and done with John's case, it turned out that his propensity for cutting corners, telling half-truths, and persistently putting self-interest above all other considerations had escalated over time from unethical behavior to criminal behavior. Ironically, the same characteristics, behaviors, and choices that propelled John to the top in record time became what eventually led to his ignominious downfall. If not always, this is at least often the case with unethical people. Later the "Johns" of the world typically say the same thing: "I didn't think I would get caught."

If this were a made-for-television movie, there would be a postscript with the good news that after John's well-publicized departure Joe was hired as the company's new CEO to clean up the mess his former friend had made. There would be poetic justice in such an ending, but that's not what really happened. In reality, Joe continued his slow but steady climb up the career ladder. Eventually, he did make it to the top in his profession and become a CEO. It took him longer to get to the top than John, but when he got there he stayed. And, during his climb to the top, because of his integrity, Joe was able to sleep well at night on that most comfortable of pillows—a clear conscience.

CHARACTERISTICS OF PEOPLE WITH INTEGRITY

People with integrity are said to have an internal moral compass. In all situations, they are guided by this moral compass rather than the ever-fluctuating whims of peer pressure, circumstances, and other influences. People of integrity behave the same way when among friends or strangers, when out of town or at home, when in a crowd or alone, and when observed or unseen. What follows are some personal characteristics common to people of integrity:

- *Consistent sameness.*   With people of integrity there is neither pretense nor hypocrisy. They are who they are. People of integrity refuse to flow with the ever-changing currents of social trends, go with the crowd, or submit to peer pressure when making decisions.

- *Value-driven.*   Their personal value system is so much a part of people of integrity that the person and values are inseparable. All aspects of their lives are guided by their internal moral compass. Because of their values, people of integrity admit when they make mistakes or are responsible for the errors of others. In addition, they behave as if they are always being observed by the ever-present eye of their moral compass—which they are.

SUCCESS TIP

A leader without followers is no leader at all. This is why integrity is so important to those who aspire to leadership positions in their field. A leader can lead only if people will follow, and people will follow only if they believe in, trust, and respect the leader.

■ *Consistency in behavior.*   Because their behavior is guided by their personal value system, people of integrity are consistent and predictable in their behavior. They are the same persons when out of town that they are when at home. They can be depended on to keep their word and to persist in doing what is right rather than what is expedient.

■ *Consistency and fairness in decisions.*   People of integrity make their decisions based on what is best for their organizations or for the greater good rather than self-interest. Where less-ethical people will try to cloud the issues and talk about "gray areas," people of integrity use their internal moral compass to see through the fog that can be generated by misguided self-interest. In doing so, they are able to find the ethical choice.

■ *Concerned more with substance than image.*   One of the downsides of the age of television is a lopsided concentration on image. For example, it is believed by political science professionals that Abraham Lincoln, who is acknowledged as one of America's best presidents but who was gangly and homely in appearance, could not be elected president in today's image-conscious age. There is nothing wrong with being concerned about your image—you should be. Presenting a positive professional image will be an important strategy in building a winning career, and is therefore the subject of Chapter 8. But to focus on image to the exclusion of substance is a mistake. An empty bag is still an empty bag no matter how well decorated on the outside.

■ *Selfless in human interaction.*   One cannot be a self-centered, self-serving person when dealing with others and also be a person of integrity. People who are ultimately the most successful are the ones who are good stewards of the resources for which they are responsible—human, financial, and physical. Selflessness means that you think first of others before thinking of yourself. Selfless people care about the greater good first.

WHY INTEGRITY IS SO CRITICAL IN BUILDING A WINNING CAREER

A leader without followers is no leader at all. This is why integrity is so important to those who aspire to leadership positions in their field. A leader can lead only if people will follow, and people will follow only if they believe in, trust, and respect the leader. Belief, trust, and respect are all dependent on the integrity of the leader. Without these characteristics, a leader will be without followers. Here are some concrete reasons why integrity is so important to career success:[5]

- *Integrity builds trust.*   Most of your work in engineering will be done in teams. Team members want to work with other team members they can count on and trust. In addition, people are more likely to follow someone they trust. On the other hand, people will only reluctantly and half-heartedly *go along* with leaders and coworkers they distrust. People who are trusted are more likely to get action from their coworkers, colleagues, teammates, and direct reports. People who are not trusted tend to get lip service, half-hearted compliance, and tactical disobedience (people pretending to comply when, in fact, they aren't).
- *Integrity leads to influence.*   In order to lead people you must be able to influence them sufficiently to gain their commitment. People will not be influenced by those they cannot trust. People who fail to win the trust of others will not have the influence on their teammates that is needed to achieve buy-in, commitment, and a willingness to persevere when the job becomes difficult.
- *Integrity establishes high standards.*   People with integrity consistently set a positive example of doing the right thing in terms of both the organization's goals and the ethical standards as opposed to taking a more expeditious or self-serving path. They say "Do as I do" rather than "Do as I say, not as I do." By their positive examples, people with integrity set high standards for their teammates, and high standards are necessary to success in today's competitive marketplace.
- *Integrity establishes a solid foundation rather than just an image.*   Many people who want to be successful spend a great deal of time, money, and effort trying to look successful. They *dress for success,* hire personal trainers, spend hours in the gym, and enroll themselves in image-building seminars. While image can be an important factor in success, image without substance will eventually lead to failure. People who lack integrity will eventually show it regardless of how hard they work to maintain an image of success. For those who want to build a winning career, a reputation for integrity must come first, and image second. The good news is that, in the long run, if you establish a reputation for integrity, projecting a positive professional image becomes much easier.
- *Integrity builds credibility.*   Credibility is what you have when others believe in and respect you, and it is a critical characteristic for those who are building a career. In order to establish credibility, there must first be trust. People will neither believe in nor respect someone they do not trust. If you want to establish credibility, first establish trust.

ETHICAL DIMENSION OF INTEGRITY

Ethics is the everyday application of a value system. It follows then that *ethical behavior* means doing the right thing as *right* is defined within a given value system. In order to behave ethically, you must have an established framework—a set of internalized values that guide your behavior, decisions, and everyday interactions with other people. In order

SUCCESS TIP

The most difficult foe of the person who wants to do the right thing instead of the popular or expedient thing is that most seductive and tenacious enemy of integrity—your own self-interest.

to succeed in the long run, people must set a consistent and positive example of ethical behavior, and in order to set such an example they must have an internal moral compass that guides them.

People in engineering careers often face ethical dilemmas in which they must balance their self-interests with the needs of the organization, make decisions that can affect the organization's profitability in both the short and the long run, balance their personal responsibilities to family and work, and withstand both overt and covert pressure to cut corners in the name of short-term profitability. Making decisions that have high ethical content can cause people to undergo what is commonly known as *soul searching.* This typically amounts to weighing what you truly believe is right against the various pressures you feel when making decisions, while also factoring in the potential personal consequences of deciding one way or the other.

Making unpopular decisions is one of the most difficult and most frequently faced dilemmas of people who rise to leadership positions. This is because in order to make a decision that is *right* as opposed to *popular,* you must have the moral courage and resolve to persevere against the inevitable pressure that will come from a variety of sources (e.g., the marketplace, supervisors, colleagues, subordinates, and sometimes even family members). But the most difficult foe of the person who wants to do the right thing instead of the popular or expedient thing is that most seductive and tenacious enemy of integrity—your own self-interest. Most human beings are driven by self-interest. Every time you make an unpopular decision, you put your self-interest at risk. Even when they know you are right, people will rarely thank you for making a decision that runs counter to their desires because they, like most people, are driven by self-interest.

Consequently, you should never be so naïve as to expect those opposed to your decisions to just sit back and willingly accept their fate. If the person who is unhappy with your decision is a customer, he might retaliate by taking his business elsewhere while pointedly blaming this unwelcome action on you. You might then find your job security and economic self-interest threatened because your ethically correct decision caused your company to lose an important customer.

If the unhappy person is your supervisor, doing the right thing might mean putting at risk that raise you have worked so hard to earn. It might even threaten promotions. If the unhappy person is a colleague, your moral stand might cause you to be isolated from the group—shunned personally and professionally. If the unhappy people are your direct reports, doing the right thing might undermine the camaraderie you worked

SUCCESS TIP

No person comes to understand the sometimes-harsh unfairness of life better than those who are successful enough to become top-level decision makers. When leaders are right, they are *heroes*, and people are quick to forget the self-interested criticism they aimed at the leader for making what they—the critics—thought was the wrong decision. But when leaders are wrong, they are *goats*, and circumstances, justifications, and logic rarely matter to their detractors.

so hard to establish in the team and replace it with an attitude of reluctant compliance or even tactical disobedience.

What these few examples say collectively is that people sometimes pay a price in personal consequences for doing the right thing—at least in the short term. At the very least, the threat is always there. This fact alone prevents many people who would otherwise be successful executives from ever rising to that level. Remember, the payoff for integrity is often delayed until sometime in the future.

If the pressure on you to decide in their favor came only from self-interested teammates, integrity would be only half as difficult a concept as it is. However, even while receiving short-term pressure from various sources when making decisions, you will also be pressured by consideration of the potential long-term effects of your decisions. Leaders who know what is right but succumb to the predictable pressures of the moment do not have the luxury of saying "I told you so" when the long-term consequences of their decisions turn out badly for stakeholders. This is perhaps the most bitter irony universally faced by people climbing the career ladder, regardless of their field of endeavor. Those who for reasons of self-interest pressure you most persistently to choose Option "A" will be first in line to criticize when the long-term results show that you—in spite of their pressure—should have chosen Option "B." Newspapers and professional journals are replete with stories that chronicle the startling downfalls of companies whose executives let themselves be pressured into making unethical decisions that served the personal interests of some in the short term, but had a devastating effect on the company in the long run.

The irony of the *double jeopardy* faced by people who make decisions in organizations might seem unfair to you, and it probably is. But, then, this is just one of the many reasons why so few people ever rise to the top in their fields. No person comes to understand the sometimes-harsh unfairness of life better than those who are successful enough to become top-level decision makers. When leaders are right, they are *heroes*, and people are quick to forget the self-interested criticism they aimed at the leader for making what they—the critics—thought was the wrong decision. But when leaders are wrong, they are *goats*, and circumstances, justifications, and logic rarely matter to their detractors. This is the origin of the old saying *It's truly lonely at the top*.

MAKING ETHICAL DECISIONS

People at work sometimes struggle with deciding what the right thing to do actually is in a given situation. In some cases, the struggle is the result of an individual's attempts to rationalize a decision he knows is unethical, but wants to make anyway for reasons of self-interest. However, there are times when even good, ethical people will be stumped—they simply won't know what the right thing to do is in a given situation. If you ever find yourself in this position, just ask the following question:

■ How would I feel about my decision in this matter if all the details about it were printed on the front page of my local newspaper to be read by my family, friends, colleagues, and teammates?

This exercise will usually clear away any mental or ethical fog surrounding the issue and get you focused on the right course of action.

A FINAL WORD ABOUT INTEGRITY

No human being is likely to ever achieve perfect integrity. It is important to understand this fact so as not to become overwhelmed by the magnitude of the challenge. But, on the other hand, neither can you afford to let *nobody's perfect* become an excuse to stop trying to live a life of integrity. While it is true that you are not likely to achieve perfect integrity, it is equally true that the more persistently you try, the closer you will come, and the closer you come, the more successful you will be in the long run. When you find yourself struggling with ethical questions, and the pressure of the moment causes you to look for ways to rationalize making the wrong decision, remember this unalterable truth:

There is no right way to do a wrong thing.

 SUCCESS PROFILE

Paul S. Hsu, PhD
Engineering Entrepreneur

Today Dr. Paul Hsu (pronounced "Shoe") is one of the most successful small-business executives in the United States. (Until recently he was chairman and CEO of Manufacturing Technology, Inc. (MTI), a company he founded himself after beginning his career as an engineer for Harris Corporation. MTI is an aerospace and defense contractor located in Fort Walton Beach, Florida, that employs more than 500 people and is growing steadily. But, as successful as Hsu is now, there was a

(*continued*)

(continued)

time when things were much different—when he was just a bright and determined young man from Taiwan equipped with nothing more than a dream; a dream that meant that in spite of language, cultural, and financial barriers he would need to emigrate to the United States and pursue a college education.

Although it was bumpy in the beginning, Hsu's road to success began in a college classroom. While going to college, a lack of financial resources sometimes presented him with tough choices: Should I buy food or should I buy textbooks? But Hsu persevered, completed his college studies, and began what has been and still is an exemplary career.

In building the winning career he now enjoys, Hsu applied many of the strategies recommended in this book, and he continues to do so even now. He is a person of impeccable integrity with a *can-do* attitude who has become a widely recognized leader in his field. Even before beginning his college studies, Hsu established high expectations for himself and developed a plan for achieving them. Part of that plan was to leave the comfortable surroundings of home, relocate to the United States, and enroll in college. Undeterred by the obstacles—and there were many—Hsu began his career journey and persevered until he succeeded.

Hsu is now recognized in the defense industry as one of the most dynamic and entrepreneurial engineering executives in the United States. Just a small sample of his many awards illustrates how far Hsu has come in his career since graduating from college. The long list of his awards includes the most prestigious of those given to small businesses in the United States: the U.S. government's prestigious National Small Business Prime Contractor of the Year; the U.S. Small Business Administration's Small Business Person of the Year; the Department of Defense's "Quality Vendor" award; and the Ernst & Young Entrepreneur of the Year.

Not only is Hsu a successful small business leader at the local level; he is also having a positive impact on small businesses at the national and international levels. Hsu serves on the President's Export Council (PEC), the membership of which includes the CEOs of some of the largest corporations in the United States, members of Congress, and the U.S. Secretaries of State, Commerce, Treasury, Labor, and Agriculture. The PEC is the premier national advisory council to the President of the United States on export issues. Hsu has accompanied the President of the United States and other dignitaries, government officials, and business leaders on trade missions around the world. On one such mission to China he personally interacted with that country's top government leader.

By any standard, Hsu is enjoying a successful career in his field. He has come a long way since his days as a college student sometimes forced to choose between books and food. In addition to the other strategies Hsu applied in building his career, one in particular stands out—his willingness to help other small businesses succeed through mentoring.

Hsu now devotes a great deal of time and effort to helping small businesses around the world succeed. His leadership in the Department of Defense Mentor-Protégé Program and the Boeing Supplier Advisory Council has resulted in expanded opportunities for numerous small businesses. His active participation as the small business representative on the Contract Innovation Group of the Air Force Materiel Command and the Air Force Acquisition Reform Leadership Council continues to pave the way for other small firms that want to do business with the U.S. Air Force.

Hsu's work as a member of the U.S. Small Business Administration's National Advisory Council has helped to ensure that small firms are not denied opportunities for government contracts simply because of burdensome bureaucratic barriers. Some of Hsu's greatest contributions in helping other small businesses have come from his work on the Regulatory Enforcement Fairness Board, which serves as an independent source of advice to the U.S. Congress concerning regulatory fairness as it relates to small businesses.

Small businesses that want to pursue Department of Defense contracts often find themselves bogged down under volumes of government regulations. Because small businesses cannot afford to employ entire departments dedicated to navigating safely through the rocks and shoals of government regulations, they are sometimes at a disadvantage when it comes to pursuing competitive contracts. Through his work on the Regulatory Enforcement Fairness Board, Hsu helps ensure that small businesses are not punished for simply being small and that they are given truly fair and equitable opportunities to compete.

Dr. Paul Hsu exemplifies the various success strategies advocated in this book. Using these strategies, he has transformed himself from a hungry college student persevering to overcome a long list of obstacles into a successful engineering executive who is having a positive impact on small businesses worldwide. His is an example every college student would do well to follow.

The ultimate evidence of his success as an engineer and entrepreneur occurred after Hsu had steadily built his company up over a twenty-year period. He began to receive offers to sell MTI, and the amounts offered were staggering. After twenty years, Hsu was ready to move on to other challenges, but as a person of integrity he made it clear to all potential buyers that a bottom-line condition for selling his company would be job security for his employees. Many potential buyers approached Hsu with enticing offers, but unless they would guarantee that his employees would be protected he sent them away. Finally, a buyer approached Hsu with an excellent offer and a guarantee of job security for his employees. When the ink dried on the sales contract, Dr. Paul Hsu, a person who once had been forced to choose between textbooks and food, was holding a check made out to him for $75 million.

REVIEW QUESTIONS

1. Use your own words to define the term *integrity*.
2. Why do you think many people do not live lives of integrity?
3. Has cheating become an accepted way of life in our society?
4. List and explain the characteristics of people with integrity.
5. Why is integrity so critical in building a winning career?
6. What is meant by the term *ethics*?

7. Explain how to make a decision that has critical ethical implications.
8. Might a person who lives a life of integrity ever make a decision she is not proud of or that is knowingly the wrong decision?

DISCUSSION QUESTIONS

1. Kathy faces a real dilemma. She is director of the design and engineering department at XYZ, Inc. One of her direct reports, a CAD technician named Don, is adding hours he is not working to his timesheet each week and being paid overtime wages for the extra hours. Kathy just found this out. On the one hand, she wants to confront Don and make him correct his timesheets and stop this unethical practice. On the other hand, Don has already been paid the extra wages for several weeks. If the Human Resources Department learns about what Don has been doing, he will be fired on the spot. Kathy likes Don and does not want him to lose his job, but she is very uncomfortable letting him knowingly turn in false timesheets—timesheets she has to sign as being correct and accurate. How should Kathy handle this situation?

2. As the Engineering Department Manager, Mike has to select one of two employees for a promotion to a job that has opened up in his department. Wanda is not as qualified for the position as Susan, but Mike likes Wanda. In fact, he would really like to get to know her better in a social setting. Mike doesn't like Susan. She is better qualified than Wanda, but she is all business. She refuses to joke around with Mike and others in the department. Which person should Mike recommend for the promotion and why?

3. Jane sometimes wishes she had never been promoted to supervisor in her company. Her boss at ABC Engineering, Mack Dunne, demands peak performance from supervisors such as Jane who lead the various departments in the company, and he does not care what they have to do to achieve it. Jane is married to a manager in another company that is competing with ABC for an important government contract. Mack Dunne is pressuring Jane to pry information out of her husband that will help ABC turn in a more competitive bid for the contract. He has hinted to Jane that if ABC wins this contract, she might receive a major promotion and salary increase. On the other hand, if ABC loses the bid the company will probably have to lay off employees including some supervisors. Jane finds herself in a real dilemma. What should she do?

ENDNOTES

1. A. Gostick and D. Telford, *The Integrity Advantage* (Layton, UT: Gibbs Smith, 2003), xii.
2. D. Callahan, *The Cheating Culture* (Orlando, FL: Harcourt, Inc., 2004), viii.
3. Ibid., 8–12.
4. Ibid., 20–24.
5. John C. Maxwell, *Developing the Leader Within You* (Nashville, TN: Thomas Nelson, Inc., 1993), 38–44.

CHAPTER TWO

Become an Effective Communicator

It's still embarrassing. I asked my caddie for a sand wedge, and ten minutes later he came back with a ham on rye.[1]

   Chi Chi Rodriguez

Few things will contribute more to the success of your career than the ability to communicate effectively. Communication skills will help you win the commitment of others to projects for which you are responsible; understand what others are thinking and feeling; sell your ideas to superiors, colleagues, direct reports, customers, and suppliers; and help your team members understand the big picture as well as where they fit into it. Good communication is critical to your success and to the success of your organization. In fact, if the workplace were a machine, communication would be the oil that keeps it running smoothly.

   Of all the skills needed in the workplace, communication may be the most important. Communication is fundamental to leadership, motivation, problem solving, ethics, discipline, training, mentoring, and all other areas that can enhance your career success. This chapter explains how to develop the communication skills you will need in order to build a winning career.

DEVELOPING COMMUNICATION SKILLS

Communication may be the most imperfect of all human processes. This is because the quality of communication is affected by so many different factors (e.g., speaking ability; hearing ability; language barriers; differing perceptions or meanings based on age, gender, race, nationality, and culture; attitudes; nonverbal cues; and level of trust between sender and receiver). Because of these and other factors, effective communication is difficult at best. However, it is essential to career success. Communication skills can be learned, which is fortunate for engineering professionals. With sufficient training and practice, most people—regardless of their innate capabilities—can learn to communicate well.

   An individual I will call Molly learned about the importance of effective communication while still a sophomore at college. Having missed the first class in an important course, Molly asked a friend to fill her in on what she had missed. The friend obliged, but there was a major error in their communication. The friend told Molly she was supposed to read ten chapters in the textbook and answer all of the essay questions at the end of the chapters. The answers to the essay questions were to be typed and submitted for a grade.

Molly thought this sounded like an unusually large assignment for one class meeting, but rather than ask clarifying questions of her friend, Molly assumed she had heard right and got started on the assignment. With such a huge assignment, Molly was glad the class met only once a week. This gave her six days to complete the assignment, and she would need every one of them.

Completing so large an assignment on time would mean that Molly had to cancel a date, an invitation to a party, and a weekend visit to a friend's home; all of which she did, but with great reluctance. When the day of her next class meeting arrived, a sleep-deprived, bleary-eyed Molly had completed the marathon assignment just barely in time to stumble zombielike into class. After turning in her essays, Molly dozed off and slept through most of the professor's lecture. It wasn't until the professor returned her essays that Molly realized her mistake. She had completed, in just one week, all of the assignments that were due at mid-term.

Rather than assuming she had heard correctly when her friend explained the assignment, Molly could have made sure by asking a few simple questions. Failing to do so cost Molly a date, a party, a weekend at a friend's house, and several sleepless nights. Molly learned an important lesson about communication as a result of this episode. From that point forward, Molly never again assumed. Rather, she always asked clarifying questions to verify what she thought she heard others say.

COMMUNICATION DEFINED

Inexperienced engineering professionals sometimes confuse *telling* with *communicating*. When a problem develops they are likely to protest, "But I told him what to do." Inexperienced engineering professionals also occasionally confuse *hearing* with *listening*. They are likely to say, "That isn't what I told you to do. I know you heard me. You were standing right next to me!"

In both cases, the individual in question has confused telling and hearing with communicating. What you say is not necessarily what the other person hears, and what the other person hears is not necessarily what you intended to say. The key word here is "understand." Communication may involve telling, but it is not *just* telling. It may involve hearing, but it is not *just* hearing. I define communication as follows:

> Communication is the transfer of information that is received and fully understood from one source to another.

A message can be sent by one person and received by another, but until the message is fully understood, there is no communication. This applies to spoken, written, and non-verbal messages.

SUCCESS TIP

Effective communication is a higher level of communication because it implies understanding *and* acceptance. The acceptance aspect of effective communication requires persuasion, motivation, monitoring, and leadership. These are the factors that separate effective communication from just communication.

COMMUNICATION VERSUS EFFECTIVE COMMUNICATION

When the information conveyed is received and understood, communication has occurred. However, understanding, by itself, does not necessarily make for effective communication. *Effective communication* occurs when the information received and understood is accepted and acted on in the desired manner.

For example, a team leader might ask her team members to arrive at work thirty minutes early for the next week to ensure that an important project is completed on schedule. All of the team members verify that they understand both the message and the reasons behind it. Without informing the team leader, however, two team members decide they are not going to arrive early because they don't see the need. This is an example of ineffective communication. The two nonconforming employees understood the message, but did not accept it and decided to ignore it. The team leader in this case failed to achieve acceptance of the message. Consequently, the communication was ineffective.

Effective communication is a higher level of communication because it implies understanding *and* acceptance. The acceptance aspect of effective communication requires persuasion, motivation, monitoring, and leadership. These are the factors that separate effective communication from just communication.

COMMUNICATION AS A PROCESS

Communication is a process that involves several components: *sender, receiver, medium,* and the *message* itself. The sender is the originator or source of the message. The receiver is the person or group for whom the message is intended. The message is the information that is conveyed, understood, accepted, and acted on. The medium is the vehicle used to convey the message.

There are three basic categories of communication media: *verbal, nonverbal,* and *written.* Verbal communication includes face-to-face conversation, telephone conversation, speeches, public announcements, press conferences, and other means of conveying the spoken word. Nonverbal communication includes gestures, facial expressions, voice tone, body poses, and proximity. Written communication includes letters, e-mail, memoranda, billboards, bulletin boards, manuals, books, and any other means of conveying the written word.

Technological developments are significantly affecting our ability to convey information. These developments include word processing, satellite communication, computer modems, cellular telephones, answering machines, facsimile machines, pocket-sized dictation machines, e-mail, and the Internet. The better you become at using all of the various media, the more effective you will be as a communicator.

COMMON INHIBITORS OF COMMUNICATION

As advanced as communication-enhancing devices have become, there are still numerous inhibitors of effective communication. Engineering professionals should be familiar with these inhibitors and learn to avoid or overcome them.

- ■ *Differences in meaning.*   Differences in meaning are inevitable in communication because we all have different backgrounds and levels of education. We might also come from different cultures, races, or countries. As a result, the words, gestures, and facial expressions we use can have altogether different meanings. To overcome this inhibitor, you must invest the time necessary to get to know the people you work with.
- ■ *Insufficient trust.*   Insufficient trust can inhibit effective communication. If receivers don't trust senders, they may be overly sensitive or guarded. They might concentrate so hard on reading between the lines for a "hidden agenda" that they miss the real message. This is why trust building is so important.
- ■ *Information overload.*   Because of advances in communication technology and the rapid and continual proliferation of information, we often find ourselves with more information than we can process effectively. This is known as "information overload," and it can actually cause a breakdown in communication. You can guard against information overload by screening, organizing, summarizing, and simplifying the information you convey to others.
- ■ *Interference.*   Interference is any external distraction that prevents effective communication. This might be something as simple as background noise caused by people talking or as complex as atmospheric interference with satellite communications. Regardless of its nature, interference either distorts or completely blocks the message. You must be attentive to the environment when trying to communicate with others.
- ■ *Condescending tones.*   Problems created by condescension result from the tone rather than the content of the message. People do not like to be talked down to, something you should never do.
- ■ *Listening problems.*   Listening problems are one of the most serious inhibitors of effective communication. They can result from both the sender not listening to the receiver and vice versa. To be a good communicator, you must be a good listener.
- ■ *Premature judgments.*   Premature judgments by either the sender or the receiver can inhibit effective communication. This is a type of listening problem because as soon as we make a quick judgment, we are prone to stop listening. One cannot make premature judgments and maintain an open mind. Therefore, it is important to listen nonjudgmentally when communicating with employees.
- ■ *Inaccurate assumptions.*   Our perceptions are influenced by our assumptions. Consequently, inaccurate assumptions can lead to inaccurate perceptions. Here's an example. Jane Jones, a technician, has been taking off an inordinate amount of time

from work lately. Her team leader Mary Andrews assumes Jane is goldbricking. As a result, whenever Jane makes a suggestion in a team meeting, Mary assumes she is just lazy and suggesting the easy way out. It turns out that Mary's assumption is wrong. Jane is actually a highly motivated, highly skilled worker. Her excessive time off is the result of a problem she is having at home, a problem she is too embarrassed to discuss. Because of an inaccurate assumption, Mary is missing out on the suggestions of a highly motivated, highly skilled team member. In addition, her misperception points to a need for building trust. Perhaps if Jane trusted Mary more, she would be less embarrassed to discuss this personal problem with her.

■ *Technological glitches.* Software bugs, computer viruses, dead batteries, power outages, and software conversion problems are just a few of the technological glitches that can interfere with communication. The more dependent we become on technology for conveying messages, the more often these glitches will interfere with and inhibit effective communication.

LISTENING AS A COMMUNICATION TOOL

Hearing is a natural process, but listening is not. A person with highly sensitive hearing abilities can be a poor listener. Conversely, a person with impaired hearing can be an excellent listener. Hearing is the physiological process of decoding sound waves, but effective listening requires perception. I define listening as follows:

> Listening is receiving a message, correctly decoding it, and accurately perceiving what is meant by it.

Inhibitors of Effective Listening

Listening breaks down when the receiver does not accurately perceive the message. Several inhibitors can cause this to happen:

■ Lack of concentration
■ Preconceived notions
■ Thinking ahead
■ Interruptions
■ Tuning out

SUCCESS TIP

One of the keys to understanding nonverbal cues lies in the concept of *consistency*. Are the spoken message and the nonverbal message consistent with each other? They should be.

To perceive a message accurately, you must concentrate on what is being said and how it is being said. Another part of effective listening is properly interpreting nonverbal cues (covered in the next section).

Concentration requires you to eliminate as many extraneous distractions as possible and to mentally shut out the rest. *Preconceived notions* can cause you to make premature judgments that turn out to be wrong. Be patient, wait, and listen. People who jump ahead to where they think the conversation is going may get there only to find themselves all alone. *Thinking ahead* is typically a response to being hurried, but it takes less time to hear someone out than it does to start over after jumping ahead in the wrong direction.

Interruptions not only inhibit effective listening, they also frustrate and often confuse the speaker. If clarification is needed during a conversation, make a mental note of it and wait for the speaker to reach a stopping point. Mental notes are preferable to written notes. The act of writing can distract the speaker or cause you to miss the point. If you find it necessary to make written notes, keep them short.

Tuning out also inhibits effective listening. Some people become skilled at using body language that makes it appear they are listening when in fact their mind is focused elsewhere. You should avoid the temptation to engage in such ploys. An astute speaker may ask you to repeat what he or she just said.

You can improve your listening markedly by applying the following strategies:

- Remove all distractions.
- Put the speaker at ease.
- Look directly at the speaker.
- Concentrate on what is being said.
- Watch for nonverbal cues.
- Make note of the speaker's tone.
- Be patient and wait.
- Ask clarifying questions.
- Paraphrase and repeat what the speaker has said.
- Control your emotions.

Listening as a success strategy is covered in more detail in Chapter 17.

NONVERBAL COMMUNICATION

Nonverbal messages represent one of the least understood but most powerful modes of communication. Nonverbal messages are often more telling than verbal ones, provided you are attentive and able to read them. Nonverbal communication is sometimes called "body language," a partially accurate characterization. Nonverbal communication does include body language, but body language is only part of nonverbal communication. There are actually three components to nonverbal communication: body factors, voice factors, and proximity factors.

Body Factors

A person's posture, poses, facial expressions, gestures, and dress—body factors—can convey a variety of messages. Even such extras as makeup or the lack of it, well-groomed or unkempt hair, and clean or scruffy shoes can convey a message. You should be attentive to these body factors and how they add to or detract from your verbal message. Do your body factors say "I am a winner" or do they say "I am a loser"?

One of the keys to understanding nonverbal cues lies in the concept of *consistency*. Are the spoken message and the nonverbal message consistent with each other? They should be. To illustrate this point, consider the hypothetical example of Mec-Tech Engineering Company. An important element of the company's corporate culture is attractive, conservative dress. This is especially important for Mec-Tech's marketing team. For men and women, white shirts/blouses, dark suits, and formal shoes are the norm.

James Oden is an effective marketing representative for Mec-Tec, but lately he has taken to flashy dressing. He has begun to wear loud sport coats, open-neck print shirts, and casual shoes. When questioned by his supervisor, James said he understood the dress code and agreed with it. This is an example of inconsistency. His verbal message says one thing, but his nonverbal message says another. This is an exaggerated example; inconsistency is not always so obvious. In fact, a simple facial expression or a subtle gesture can be an indicator of inconsistency.

When verbal and nonverbal messages are inconsistent, you should dig a little deeper. An effective way to deal with inconsistency is to gently but frankly confront it. A simple statement such as "Cindy, your words say you agree with me, but your eyes say you disagree" can help draw a person out so that you get the real message.

Voice Factors

Voice factors are also important elements of nonverbal communication. In addition to listening to employees' words, you should listen for voice factors such as volume, tone, pitch, and rate of speech. These factors can reveal feelings of anger, fear, impatience, uncertainty, interest, acceptance, confidence, etc.

As with body factors, it is important to look for consistency when making note of voice factors. It is also advisable to look for groups of nonverbal cues. A single cue taken out of context has little meaning. But as one of a group of cues, it can take on significance. For example, if you look through an office window and see a person pounding her fist on the desk, it would be tempting to interpret this as a gesture of anger. But is it really? What kind of look does she have on her face? Is her facial expression consistent with desk-pounding anger, or could she just be trying to open a drawer that is stuck? On the other hand, if you saw her pounding the desk with a frown on her face and heard her yelling in an agitated tone, your assumption of anger would be well based. She might be angry because her desk drawer is stuck; nonetheless, she would still be angry.

Proximity Factors

Proximity factors range from where you position yourself when talking with someone to how your office is arranged, the color of the walls, and the types of fixtures and decorations you have. A supervisor who sits next to an employee during a conversation conveys a different message than one who sits across the desk from the employee. A supervisor who makes his or her office a comfortable place to visit is sending a message that invites communication. A supervisor who maintains a stark, impersonal office sends the opposite message. To send the nonverbal message that people are welcome to stop by and talk, consider the following suggestions:

- Have comfortable chairs available for visitors.
- Arrange chairs so you can sit beside visitors rather than behind your desk.
- Choose soft, soothing colors rather than harsh, stark, or overly bright or busy colors.
- If possible, have refreshments such as coffee, soda, and snacks available for visitors.

VERBAL COMMUNICATION

Verbal communication ranks close in importance to listening. You can improve your verbal communication skills by being attentive to the following factors:

- *Interest.*   When speaking with people, show an interest in the topic. Show that you are sincerely interested in communicating your message to them. Demonstrate interest in the receivers of the message, as well. Look them in the eye or, if in a group, spread your eye contact evenly among all receivers. If you seem bored, reluctant, or indifferent, people will notice and follow your example.
- *Attitude.*   A positive, friendly attitude enhances verbal communication. A caustic, superior, condescending, disinterested, or argumentative attitude will shut off communication. Be patient, be friendly, and smile.
- *Flexibility.*   Be flexible. For example, if you call your team members together to explain a new company policy but find that they are uniformly focused on a problem that is disrupting their work schedule, be flexible enough to put your message aside for now and deal with the current problem. Until your team members deal with what's on their minds, they will not be good listeners.
- *Tact.*   Tact is an important ingredient in verbal communication, particularly when delivering a sensitive or potentially controversial message. Tact has been referred to as the ability to hammer in the nail without breaking the board. The key to tactful verbal communication is thinking before talking.
- *Courtesy.*   Being courteous means showing appropriate concern for the needs of the receiver. Calling a meeting ten minutes before the end of the workday, for example, is discourteous and will inhibit communication. Courtesy also dictates that you avoid monopolizing. When communicating verbally, give the receiver ample opportunities to seek clarification and to state her point of view.

COMMUNICATING CORRECTIVE FEEDBACK

When you become a team leader or supervisor, you will occasionally need to give corrective feedback to employees. Corrective feedback, if effectively given, will help them improve their performance. To be effective, however, corrective feedback must be communicated properly. The following guidelines will help enhance the effectiveness of corrective feedback:[2]

- *Be positive.* To be corrective, feedback must be accepted and acted on by the employee. This is most likely to happen if it is delivered in a positive manner. Give the employee the necessary corrective feedback, but don't focus only on the negative. Find something positive to say.
- *Be prepared.* Focus on facts. Do not discuss personality traits. Give specific examples of the behavior you would like to see corrected.
- *Be realistic.* Make sure the behaviors you want to change are within the control of the employee. Don't expect an employee to correct a behavior she does not control. Tell the employee about her behavior, ask for her input, and listen carefully when that input is given.

REVIEW QUESTIONS

1. Are people born good communicators, or can they learn to be good communicators? Explain.
2. Distinguish between communication and effective communication.
3. List and explain the various components of communication.
4. List and explain common inhibitors of communication.
5. Define in your own words the term *listening*.
6. List and explain common inhibitors of effective listening.
7. Explain ten strategies you can use for improving your listening.
8. List and explain the three components of nonverbal communication.
9. Explain how you can improve your own verbal communication.
10. List and explain three strategies for enhancing the effectiveness of corrective feedback.

DISCUSSION QUESTIONS

1. John Adams is having communication problems in his team. When he first became the team leader, he simply told his team members what he wanted them to do and when they had to complete their assignments. But he soon noticed that team members tended to misinterpret his instructions. Consequently, of late he asks team members to repeat his instructions to make sure they have heard him and understood. This approach cut down on the communication problems in his team somewhat, but did not eliminate them. For example, one morning he told a team member to complete a project by noon. When at noon the employee had not brought the completed work to him, Adams stopped by his cubicle to inquire about the project only to find that his tardy team member had gone to lunch. "I don't get

it," mumbled Adams to himself. "I told him I needed that project completed by noon and I know he heard me and understood what I said." If John Adams were a friend of yours, what advice would you give him to improve communication in his team?

2. Phyllis Appleton's team members complain constantly that she does not listen to them. "Sure," they say, "her door is always open when we need to talk. But that does not mean she listens. She interrupts constantly, jumps ahead when she thinks she knows where you are going, and continues to do paperwork instead of concentrating on what we say." Is Phyllis Appleton a good listener? Would you like to be a member of her team? What advice would you give Appleton about listening to her team members?

3. Thomas Burr's office is stark, functional, and cold. The walls are grey, the chairs are uncomfortable, and the overall feel is that of an interrogation room in a police station. What nonverbal messages does Burr's office give to people who visit him there? What kinds of problems might Burr's stark, cold office cause for him at work?

ENDNOTES

1. Louis E. Boone, *Quotable Business*, 2nd ed. (New York: Random House, Inc., 1999), 59.
2. R. A. Luke, "How To Give Corrective Feedback to Employees," *Supervisory Management*, March 1980, 7.

CHAPTER THREE

Become a Leader in Your Organization

Leadership appears to be the art of getting others to want to do something you are convinced should be done.[1]

   Vance Packard

Leadership is an intangible concept that can produce tangible results. It has been referred to by some as an art and by others as a science. In reality, leadership is both an art and a science. Organizations that are well led, whether they are small teams or large corporations, typically have better productivity, quality, financial performance, market share, and overall competitiveness. Few things will advance your career more effectively than becoming a leader in your organization and among your peers.

WHAT IS LEADERSHIP?

Leadership has been examined from many different perspectives: the military, education, business, industry, government, sports, and other fields of endeavor. Consequently, there are different definitions available for the concept. My definition attempts to encompass all fields. It is as follows:

> Leadership is the ability to inspire others to make a total, willing, and voluntary commitment to accomplishing or exceeding the organization's goals.

On the surface this definition, because of its brevity, may appear overly simple. But upon closer scrutiny you will find that it has more depth than is at first apparent. To begin, there is the element of "inspiring others." Most other definitions use the term *motivating* rather than *inspiring*. I use *inspiring* because it represents a higher-level concept than *motivating*. We motivate people by showing them the goals, providing appropriate incentives for meeting them, and giving them a continual stream of encouragement. *External motivation* is a fickle and fleeting concept that must be continually reinforced. It is like the gas tank in your car: You just have to fill it up again and again.

Inspiration, on the other hand, if not a permanent concept is at least longer lasting than motivation. We inspire people by exemplifying personal and professional characteristics they admire and would like to have themselves. People are inspired by us when they observe traits and behaviors that make them want to be like us. Think about people who inspire you, and ask yourself why. What is it about these people that inspires you? We inspire others by being good at our jobs, exemplifying integrity, being able to make difficult decisions, doing the right thing in any given situation, helping others, being selfless, being fair and consistent, and by generally being what others would like to be.

The other significant element in this definition of leadership is "total, willing, and voluntary commitment." It is easier to lead people when they want to go where you are leading; when the organization's goals become their goals; when it is just as important to them as it is to you that these goals be accomplished. This is what is meant by "total, willing, and voluntary commitment." When those you lead are as determined to achieve the organization's goals as you are, you win, they win, and the organization wins. If you become the kind of leader who can inspire others to make a total, willing, and voluntary commitment, a successful career is virtually assured for you.

CHARACTERISTICS OF GOOD LEADERS

Good leaders come in all shapes, sizes, genders, races, political persuasions, and national origins. They do not necessarily look alike, talk alike, walk alike, or even work alike, but they do share some common characteristics. What leaders have in common are the following characteristics:

- ■ *Balanced commitment.* Good leaders are committed to both the job and the people who do the job. Good leaders are also committed to both their work responsibilities and their personal responsibilities. This means they are able to strike an appropriate balance between what can sometimes be conflicting responsibilities and concerns. People who are so focused on getting the job done that they are inconsiderate and insensitive to the needs of the people who do the work will ultimately fail. On the other hand, those who are so intent on the feelings and personal needs of others that they neglect the work their direct reports should be doing will also fail. People who neglect their personal responsibilities (family, friends, personal health, etc.) in order to climb up the career ladder might ultimately find themselves at the top but alone without their family, friends, or health. On the other hand, people who neglect their work in favor of family and friends will go nowhere in their career. Those who are ultimately the most successful are people who learn how to achieve an appropriate balance between their career and other responsibilities, interests, and obligations.

SUCCESS TIP

Those who are ultimately the most successful are people who learn how to achieve an appropriate balance between their career and other responsibilities, interests, and obligations.

■ *Positive role model.* In order to be a leader, you must be a positive role model. This means you must learn to exemplify those behaviors that you expect of those you want to lead. You'll never be an effective leader if you project an attitude that says, "Do as I say, not as I do." People will follow your words only if your actions are consistent with your words. If you are a team leader and expect team members to arrive at work fifteen minutes early, you should arrive at least twenty minutes early. If you expect team members to be able to discuss and debate work-related issues without becoming angry or disagreeable, you must model this type of behavior. To become a leader among your colleagues and peers, find out what attitudes and behaviors are most important in your organization and become an exemplary role model of these attitudes and behaviors.

■ *Good communication skills.* Much of leadership is communication. Good leaders are typically good communicators. They are willing, patient, assertive listeners. They are also good at clearly and succinctly expressing their ideas and wishes. They use their communication skills to establish and nurture rapport with others. They know that communication is the oil that lubricates the human processes of an organization. Consequently, they use plenty of it.

■ *Positive influence.* Effective leaders are able to influence others in a positive way. Influence is the practical application of credibility, competence, personality, authority, and other factors to move people toward a particular destination, course of action, or point of view. When credibility, competence, personality, authority, and other leadership traits are effectively applied, they become positive influence.

■ *Persuasiveness.* Effective leaders are persuasive. People don't always do what needs to be done just because it needs to be done, and they don't always follow instructions just because instructions are given. Consequently, the ability to persuade people to accept a given point of view or course of action is important. People who are persuaded of the viability, necessity, or worthiness of a given goal are much more likely to pursue it than those who are just ordered to do so. An effective strategy for persuading people to buy into a given course of action is to explain what's in it for them. Enlightened self-interest is an excellent persuader.

■ *Sense of purpose.* Effective leaders have a sense of purpose. They know where the organization needs to go and the role they play in moving the organization in that direction. People with a sense of purpose are not confused about the organization's direction or what needs to be done in order to achieve organizational goals. A strong sense of purpose makes others more willing to follow.

■ *Self-discipline.* Effective leaders develop self-discipline and use it to set a positive example for others. Through self-discipline, leaders avoid inappropriate displays of emotion such as anger, negative self-indulgence, and counterproductive responses to the daily pressures of work and life. Self-disciplined leaders set an example of handling the inevitable problems and pressures of work with equilibrium and a positive attitude. Those who do this stand out from the crowd in ways that promote success.

■ *Honesty and integrity.* The most effective leaders are trusted by others. They win the trust of others by consistently being open, honest, and forthright in all human interactions. Because they are people of integrity, they can be depended on to make decisions—even in the most difficult of circumstances—with fairness, consistency, and equity. People who are trusted typically go farther in their careers than those who cannot be trusted.

- *Credibility.*   The most effective leaders establish a high level of credibility among their colleagues, peers, and subordinates. Credibility is established by being good at what you do (professional knowledge and skills); by being consistent, fair, and impartial in human interactions; by setting a positive example; and by adhering to the same standards of performance and behavior you expect of others.

- *Common sense.*   Effective leaders have common sense and they use it. They get to the heart of what is important in a given situation and remember what is important when making decisions. People with common sense refuse to hide behind the rulebook when putting aside the book makes more sense in a given situation. They learn when to be flexible and when to be firm. People lacking common sense often create self-inflicted roadblocks to their own success.

- *Stamina.*   Effective leaders must have stamina. Frequently, they are the first to arrive at work and the last to leave. They typically work more hours than others and their hours are often more intense and pressure packed. Consequently, energy, endurance, and good health are important to those who lead. When building a career, you won't get far if you can't keep up.

- *Commitment.*   The most effective leaders are those who commit to their personal goals as well as the goals of the organization and are willing to persevere in doing what is necessary, within the bounds of ethics and good professional practices, to achieve both sets of goals. Few people have ever succeeded without commitment.

An individual I'll call Mark illustrates the importance of commitment for those who hope to become leaders in their fields. Mark was a successful engineering executive who climbed to the top in his field by developing many of the characteristics explained in this section. Of these characteristics, the most important to Mark for the development of his career was commitment.

Mark knew he wanted to be an engineer even as a high-school student, and he knew the kinds of preparation that would be required. When he graduated from high school, Mark was anxious to begin his college studies, but there were obstacles. The first obstacle Mark faced was gaining admission to the university he planned to attend. Because his high school was small and rural, it didn't offer all of the prerequisite courses engineering majors were expected to complete before applying for admission. Mark lacked the advanced Mathematics and Science courses he needed for admission. Not to be put off, Mark enrolled in a nearby community college and spent a year completing all of the prerequisite courses required by the university.

Then, just when he thought he was set for admission to the university, Mark was confronted with another obstacle. His SAT score would admit him to the university, but not to the College of Engineering. His score was fifteen points too low. This put off admission to his university program another semester while Mark completed two intensive SAT preparatory courses. When he took the SAT a second time, Mark improved sufficiently to finally gain admission to the College of Engineering. Things seemed to finally be going his way, but then, after completing just one semester in his program at the university, Mark's father died suddenly and unexpectedly. Because his mother needed financial help, Mark was forced to drop out of the university and and go to work to help support his younger brother and sister.

Still committed to his chosen career in engineering, Mark returned to the university on a part-time basis taking courses at night and through the university's distance learning program. After two years, his family's financial situation stabilized and Mark was able to resume his studies on a full-time basis. By now, Mark was such a committed, no-nonsense student that he progressed rapidly through his course work making excellent grades.

During his last semester, Mark began interviewing for jobs through the university's career center. As an honors student with a high grade point average and some actual work experience, Mark was actively recruited by several companies. Mark surprised his professors and family by accepting a position with a company that offered not the highest starting salary, but what Mark thought would be the best opportunities for rapid promotions.

Less than a week after graduation, Mark was working in his field. He now had an excellent job with a good company. Mark had placed his feet firmly on the first rung of the career ladder with every intention of climbing up that ladder as fast as possible. But his climb was interrupted by an automobile accident.

One day while driving home from work in a rainstorm, Mark lost control of his car and slammed into a tree. Mark lived, but his injuries were serious enough to keep him in the hospital for six months and in physical therapy for another six months. A less committed individual might have given up on his career at this point, but Mark didn't. While confined to a hospital bed and in physical therapy for most of a year, Mark enrolled in graduate school through his university's distance learning program. Mark completed his graduate degree and his physical therapy at the same time.

When he returned to work, Mark was determined to make up for lost time—and he did. Within two years, Mark had caught up with his contemporaries. Within two more years he had passed by his contemporaries as he climbed up the career ladder. He continued to climb steadily until he reached the top in his field and had an excellent career. Mark's commitment brought him success in spite of a series of debilitating setbacks.

HOW YOU CAN BECOME AN EFFECTIVE LEADER

The various characteristics of effective leaders presented in the previous section will help you succeed in your career. You can begin developing and applying these characteristics while still in college. In fact, college is an excellent environment for developing leadership skills. Any time you are part of a group of people that is working toward the same or a similar goal there will be opportunities to develop leadership skills. The more leadership skills you develop while in college the more prepared you will be once you secure a position in your field after graduation.

As soon as you are employed in your field, begin applying the leadership skills you developed while in college and continue to develop those skills. In addition, apply the following strategies and begin using them to become an effective leader in your organization.

Learn Your Organization's Vision and Commit to Helping Achieve It

An organization's vision is a word picture of where it wants to go and what it wants to be. For example, an engineering firm might adopt the following vision: *To be the leading provider of engineering services in the Midwest.* If you go to work for this engineering firm, your advancement will depend on how effective you are in helping the organization achieve this vision. But it's important to do more than just your part. You should also become known as a person who provides the leadership necessary to help others do their part.

Project Unquestionable Integrity and Selflessness

The foundation of effective leadership is character, and integrity is the most important aspect of character for those who want to be leaders. Most work in organizations is done in teams where the work of one individual affects the work of others. Consequently, it is important that people at work be able to trust each other. Integrity manifests itself in the following ways that build and maintain trust. People with integrity (a) refuse to pretend—they are not hypocrites; (b) have an internal moral compass—a well-defined set of values by which all of their actions and decisions are guided; (c) are consistent in their behavior and decisions; and (d) are selfless—they think of the organization and the people they lead first and themselves second.

Establish Credibility

You cannot lead people unless they are willing to follow you, and people won't follow you unless you establish credibility with them. Credibility is the ability to establish a sense of *believing in you* among other people. You establish credibility by (a) being honest, fair, and consistent when dealing with people; (b) being good at your job and continually improving your professional knowledge, skills, and involvement; and (c) setting a consistently positive example of doing well those things that are most important in making your organization a success.

Develop and Maintain a Positive "Can-Do" Attitude

There is a story about an engineering student at Stanford University who needed a job in order to help pay the cost of his education. When he noticed an ad placed by a professor for a typist, he went directly to the professor's office to apply for the job. The interview went well, and the professor offered the student the job. The student gladly accepted,

SUCCESS TIP

As you advance in your career you will increasingly face responsibilities, obstacles, challenges, and problems that are new to you. A can-do attitude will give you the inner strength needed to face the unknown and conquer it.

but said he couldn't begin work for a couple of days. When he started work several days later, the student did an excellent job. The professor was pleased and told the student so. But he was also curious. He wanted to know why the student had to wait several days before starting work. The student responded that the professor's ad was for a typist, and although he needed the job, at the time he had applied he didn't know how to type. Before he could start the job he had to borrow a typewriter and teach himself how to type. The can-do student in this story not only became a noted engineer, he eventually became president of the United States (Herbert Hoover).

This is an excellent example of a positive, can-do attitude. A can-do attitude is the outward manifestation of an inward conviction that says, "Whatever the challenge, I can and will meet it." This is important because many a time in your career you will be asked to perform tasks you don't yet know how to do. As you advance in your career you will increasingly face responsibilities, obstacles, challenges, and problems that are new to you. A can-do attitude will give you the inner strength needed to face the unknown and conquer it. Doing this is one of the ways that people stand out from the crowd and become leaders. And becoming a leader is one of the best ways to advance your career.

Develop Self-Discipline

Many people decide to diet, but few stick to it. Many people start an exercise program, but few actually work out regularly. Many people make New Year's resolutions, but few keep them up. The missing ingredient in all of these cases is self-discipline. Once committed to a goal, people need self-discipline in order to persevere in achieving it. Self-discipline is the ability to consciously take control of your personal choices, decisions, actions, and behaviors. In order to be a good leader you must develop self-discipline. Said another way, before you can lead anyone else, you must first learn to lead yourself.

Be a Creative Problem Solver and Decision Maker

Even the best-led organizations—from small teams to large corporations—have problems. A problem is any situation in which what exists does not match what is desired. Said another way, a problem is a discrepancy between the actual and the desired state of affairs. When faced with problems at work, we have to make decisions. Decision making is simply the process of selecting what we believe to be the most appropriate course of action from among the options available to us. Consistently making good decisions is critical to those who wish to be leaders and advance in their careers. Decisions are to an organization what gasoline is to an automobile. You cannot operate an organization without making decisions, and the better the decision making the better the organization runs.

One of the things that separates leaders from other decision makers is their ability to be creative when making decisions; to step out of the box and see things from different perspectives and in different lights. Creativity is the ability to be imaginative, original, and innovative. One of the best ways to promote creativity in decision making is to conduct brainstorming sessions in which even the most seemingly outlandish proposals are welcomed and discussed.

Be a Positive Change Agent

In order to survive in a competitive environment, organizations must improve continually. This means they have to change continually. Organizations that fail to change and improve continually—whether from complacency, ignorance, or conceit—are doomed to mediocrity or, worse yet, failure. One of the best ways to be a leader in an organization is to be a person who helps it change and improve continually. This means you have to become a positive change agent. Few things call more clearly for effective leadership than organizational change. It takes an effective leader to overcome the natural reluctance of both people and organizations to change, but those who can do this are more likely to enjoy successful careers. Good leaders are able to take people where they are not yet ready to go.

Be an Effective Team Builder

The work of organizations is done by teams of people. One of the fastest ways to become a leader and to advance your career is to be such a good team player that you soon become a team leader. A team is a group of people with a common goal. Experience has shown time and again that a team of people working together in a cooperative and mutually supportive way will outperform the same number of individuals all going their own way, no matter how talented the individuals happen to be. In order for people to work well together in a group, relationships must be built, a collective identity must be established, good communication must be maintained, and ground rules must be established. Doing these things is called *team building*, and good team builders are typically also good career builders.

Empower Others to Lead Themselves

The best leaders are those who help others become self-directed, self-motivated, and self-led. This is important because leaders in an organization cannot always be there to direct, motivate, and energize. The best evidence of a well-led organization is that its people work just as hard and just as well when the "boss" is away. Achieving this desirable state of affairs involves empowering people to act within well-defined parameters and encouraging them to step forward, take the initiative, and lead themselves. The willingness and ability to empower others will help propel you up the career ladder.

 ## SUCCESS TIP

In order to lead others, you must be able to manage conflict and build consensus. Few things will advance your career more than the ability to manage human conflict and build consensus among people in an organization.

Be an Effective Conflict Manager and Consensus Builder

The workplace is rife with conflict, even in well-led organizations. Conflict can be caused by limited resources, incompatible goals, role ambiguity, differences in values, differences in perspectives, communication problems, and many other factors. Conflict is a stress producer, and stress can kill more than quality, productivity, and competitiveness. It can actually kill people. Stress is a leading contributor to heart attacks, strokes, and workplace violence. In order to lead others, you must be able to manage conflict and build consensus. Few things will advance your career more than the ability to manage human conflict and build consensus among people in an organization. Conflict management is the subject of Chapter 17.

SUCCESS PROFILE

Dr. Eleanor Baum
Engineering Educator

Dr. Eleanor Baum is one of the most widely acclaimed and highly respected engineering educators in the world. She has devoted her career to promoting engineering as a career for minorities and women. In today's college environment, promoting engineering as a career for minorities and women sounds like an admirable but normal challenge. But in 1955, when Dr. Baum first enrolled as an engineering student at the City College of New York, there were very few women and minorities in the discipline. In fact, she was the first and only female engineering student in her class. As it turned out, this would be just one of many "firsts" Dr. Baum would achieve in her long and distinguished career. She was also the first woman to serve as a dean of a college of engineering and the first to serve as the president of the American Society for Engineering Education (ASEE).

Dr. Baum selected what was then considered a "male-oriented" college major as a form of rebellion—she wanted to shock her parents; which she did. But she had not been enrolled long before discovering that she not only had a knack for engineering, but actually liked it. This allowed her to persevere through the isolation and lack of either support or encouragement. As a student Baum was like an island unto herself in one of the toughest academic disciplines, but she persevered and received a Bachelor's Degree in Engineering in 1959.

When she began her career in the aerospace industry, the isolation she had endured in college continued. She was the only woman working as an engineer in the entire company. As a result, she was often assumed by others to be a secretary. Once again, Dr. Baum had to persevere through isolation, a lack of support, and both racial and gender bias. And she did so.

(continued)

(continued)

After winning a fellowship from the Polytechnic Institute of New York, Baum returned to college and completed both a Master's and a Doctorate (PhD) degree in her field. While serving as a graduate assistant, Baum taught an undergraduate class and found that she enjoyed teaching. With this realization, a star was born and the rest is history. She joined the Pratt Institute, where, once again, she was the only woman on the engineering faculty. In 1971 she was promoted to Chair of the Electrical Engineering Department and in 1984 to the position of Dean of Engineering.

As an engineering dean, Dr. Baum reached out to disadvantaged minorities and women and, in the process, created opportunities that allowed many to pursue college degrees in engineering. In 1987, Dr. Baum became Dean of the Cooper Union Albert Nerken School of Engineering and a fellow of the Society of Women Engineers (SWE). In this position she increased the number of women enrolled in engineering majors at Cooper Union from just five percent to an amazing thirty plus percent. To put this achievement in perspective, the national average is half this number.

One of her goals has been to help all young people to view engineering as a profession that helps solve society's problems and makes life better for people. These are worthy goals that women and minorities can achieve as engineers. Another of her goals has been to maintain an appropriate balance between her professional and family responsibilities, and, like her other achievements, this is a goal she has achieved. She and her husband—working as a team—have raised two daughters. In 1996, in honor of her long and distinguished career, Dr. Baum was inducted into the Technology International Hall of Fame. In building a winning career for herself, Dr. Baum also helped many others building winning careers for themselves. Hers is an example worthy of emulation. It exemplifies the success strategies of perseverance, mentoring, balancing career and family obligations, setting high expectations, and becoming a leader.

Source: "Eleanor Baum." *Notable Women Scientists.* Gale Group, 2000. Reproduced in *Biography Resource Center,* Farmington Hills, Michigan: Thomson Gale, 2005. *http://galenet.galegroup.com* (document number: K1668000026).

REVIEW QUESTIONS

1. Use your own words to define the term *leadership.*
2. List and explain the characteristics of good leaders.
3. List the strategies you can use to become an effective leader.
4. Explain how you can establish credibility with those you hope to lead.
5. What is meant by the term *can-do attitude*?
6. Why is self-discipline so important to people who want to become leaders?
7. What is meant by the term *creative problem solver*?
8. Why is it important to be a positive change agent if you want to be a leader in an organization?

9. What is meant by the term *team building*?
10. Why is it important for potential leaders to learn to be effective conflict managers?

DISCUSSION QUESTIONS

1. Select an individual in any field who *inspires* you and discuss why. This person does not have to be well known or a celebrity.
2. Miriam Eben is a very demanding supervisor. Her only concern is getting the job done. When employees on her team have problems or need time off for family responsibilities, Eben's answer is almost always, "I am not concerned about your personal problems. My job is to make sure you get your job done." How effective a team leader do you think Eben will be in the long run? Why?
3. Pete Summers is the supervisor of the design department for ABC, Inc. Summers is known as a supervisor who tells his team members, "Do as I say, not as I do." How effective a team leader do you think Summers will be in the long run? Why?
4. Amanda Barkly has just been promoted to team leader. She is an excellent engineer, but her superiors are concerned about her people skills. However, they have decided to give Barkly a chance to prove herself as a team leader. One of Barkly's biggest problems is her temper. She does well when things are going right, but she quickly loses her temper when things don't go as planned. What problems do you think Amanda Barkly will have as a team leader?

ENDNOTE

1. Louis E. Boone, *Quotable Business*, 2nd ed. (New York: Random House, Inc., 1999), 33.

CHAPTER FOUR

Set High Expectations for Yourself and Develop a Plan for Achieving Them

Luck is a matter of preparation meeting opportunity.[1]
   Oprah Winfrey

Strategic planning is the process organizations use to decide who they are, where they are going, and how they will get there. This is an important process for organizations trying to gain strategic advantages in a competitive marketplace. At some point in your career you will probably participate in a strategic planning process for the organization that employs you. Before doing that, however, you should have already developed your own personal strategic plan.

People trying to build successful careers are like organizations trying to compete in the global marketplace: they need to know who they are, where they are going, and how they will get there. In other words, strategic planning is just as important to you as an individual as it is to any organization that might employ you.

WHY HIGH EXPECTATIONS AND A GOOD PLAN ARE SO IMPORTANT

Dennis Waitley, author of the landmark book *The Psychology Of Winning*, did an excellent job of summarizing the importance of setting high expectations and developing a comprehensive plan to achieve them when he said,

> Winners in life—those one in a hundred people—are set apart from humanity by one of their most important developed traits—*positive self-direction*. They have a game plan for life. Every winner I have ever met knows where he or she is going day after day—every day. Winners are goal and role oriented . . . they are self-directed on the road to fulfillment. Success has been defined as the progressive realization of goals that are worthy of the individual.[2]

SUCCESS TIP

Successful people are proactive, assertive, and self-directed in planning and building their own careers.

SUCCESS IS A DO-IT-YOURSELF PROJECT

Successful people invariably take responsibility for doing what is necessary to succeed. This does not mean that success is achieved without the help of others; it is not. Rather, it means that successful people don't just sit back and wait for others to develop their careers for them. Successful people are proactive, assertive, and self-directed in planning and building their own careers. Of course, they accept and appreciate the help received from others, and they help others in return, but successful people know that ultimately the responsibility for building their careers rests primarily with them.

Jeffrey Fox made this point in his book *How to Become a CEO*: "Corporations don't have career plans for future presidents. It is doubtful if they have them for anybody. Your destiny and your career growth are your responsibility, no one else's. You have to know what you want. You have to design the plan to get there."[3]

ESTABLISH HIGH EXPECTATIONS FOR YOURSELF

If you want to outperform the competition, you have to set the bar high. Olympic athletes know this. They know that in order to win the gold medal, they must establish high expectations for themselves. In fact, just deciding to pursue the gold medal is setting high expectations. This is the message conveyed by the life of Aprille Ericsson-Jackson who is the first African American woman to earn a PhD in mechanical engineering from Howard University. Dr. Ericsson-Jackson says,

Shoot for the moon and even if you miss you'll still be among the stars.[4]

This message is echoed by Rick Pitino who, in his book *Success Is a Choice*, said, "When we were little kids in the playground we didn't dream of hitting singles in the first inning. We dreamed of that homerun in the bottom of the ninth to win the World Series."[5]

SUCCESS TIP

Establishing high expectations for yourself is important, but it's just the first step. In addition to high expectations, you need a plan.

The following story illustrates the importance of setting high expectations and developing a comprehensive plan for achieving them. During college, Mary was a good student with a solid "B" average, but there were plenty of students in her class at the university with better grades. Now, years later, some of those students work for Mary. How did she get to the top of her profession when her fellow students with better grades didn't? One of the reasons is that while still in college, Mary set her sights high and developed a plan. Mary not only decided she wanted to be the CEO of an engineering firm, but she also conducted the research necessary to determine what she would have to do in order to get there. Then she developed a plan based on what she had learned.

During college, when Mary would talk with her friends about career expectations, she found they fell into one of two categories: those who were so wrapped up in the college experience that they didn't even think about life after graduation, and those who hoped to succeed, but gave little or no thought to how they should go about it. By the time Mary graduated from college, she had a personal strategic plan mapped out for becoming an engineering CEO in fifteen years. She made it in twelve.

PLAN FOR SUCCESS

Establishing high expectations for yourself is important, but it's just the first step. In addition to high expectations, you need a plan. Your personal strategic plan for career development should have the following components:

1. Career vision
2. Guiding principles
3. List of your career-related strengths and weaknesses
4. Goals and strategies

Developing Your Career Vision

A career vision is the dream of what you would like to be put into words. Your personal career vision will serve as a beacon in the distance toward which you are always progressing. Like the kind of beacon found in lighthouses, it will keep you on course as you navigate through the inevitable fog, rocks, and shoals of your career. If you have a clear career vision, it is easier to stay focused when life becomes hectic, complicated, ambiguous, or overwhelming; which is often the case. When developing your career vision, ask the following types of questions:

- When I am old, retired, and looking back on my career, what is the picture I would like to see?
- If I could choose to be anything or have any position in my profession, what would it be?
- If I were asked to write a one-sentence biographical introduction of myself that summarizes my career, what would it say?

SUCCESS TIP

Developing a career vision can require hours of heart-tugging, brain-bending, soul-searching effort. It is not an easy process. Sometimes the most difficult questions you can ask yourself are, who am I and who do I want to be?

The answers to these questions will help you craft a career vision that will satisfy the criteria that all good vision statements should meet. These criteria are as follows:

- Brief, yet comprehensive
- Easily understood
- Challenging, yet attainable
- Lofty, yet tangible
- Worthy of you

Sample Career Vision Statements

Developing a career vision can require hours of heart-tugging, brain-bending, soul-searching effort. It is not an easy process. Sometimes the most difficult questions you can ask yourself are, who am I and who do I want to be? However, once you have answered the questions just recommended, writing the actual vision statement becomes a simple process. What follows are some sample career vision statements you can use in developing a statement of your own:

- I will be the chief engineer in a large design and engineering services firm.
- I will be the CEO of a technology company.
- I will be the senior partner in a successful engineering firm.
- As a safety engineer, I will be a regional director for the Occupational Safety and Health Administration.
- I will be the quality manager for a large manufacturing firm.
- I will be the CEO of my own structural engineering and construction company.

Developing a Set of Guiding Principles

The farther you advance in your career, the tougher the decisions become. What will you do when it's your responsibility to fire an employee for good cause, but that employee

SUCCESS TIP

Coming to grips with your beliefs, values, and priorities is a critical step in developing a personal strategic plan for your career.

happens to be a friend? What decision will you make when offered a major promotion that will require relocation to a community far away from family, friends, and your personal interests (e.g., say your hobby is surfing and the new job is in Kansas)? What will you do if you need to pursue a graduate degree at night in order to advance to the next level in your career, and evenings are the only time you have to spend with your spouse and children?

How you will answer questions such as these depends on what you value most in life and what your priorities are at the time. Coming to grips with your beliefs, values, and priorities is a critical step in developing a personal strategic plan for your career. Writing down your career-related beliefs, values, and priorities is how you develop your guiding principles.

Your guiding principles establish the framework within which you will pursue your vision. Each guiding principle represents a personal core value or belief that is important to you. Taken together, the guiding principles should represent your value system as it relates to your career.

Once you have developed a comprehensive set of guiding principles, making tough career-related decisions gets easier—not easy, but easier. Career decisions are seldom easy. However, with a solid framework provided by your guiding principles, you will be able to easily eliminate any options that are contrary to your values and priorities. This, in turn, will help identify those options that are right for you.

Sample Guiding Principles

Guiding principles relating to your career are deeply personal. Yours might be radically different from those of your best friend or even from those of family members. Consequently, when reading the following examples of guiding principles, remember that they are just examples. Your guiding principles should be based on your personal values and priorities.

- *Honesty and integrity.*   I will be honest, forthright, and ethical in all my dealings with others.
- *Balanced commitment.*   I will maintain an appropriate balance between my career and personal interests.
- *Health.*   I will pursue career advancement vigorously, but not to the detriment of my health.
- *Common sense.*   I will apply common sense when making decisions and interacting with people.
- *Credibility.*   I will establish credibility in my field by mastering the knowledge and skills of my profession, and continuing to learn and grow on a lifelong basis.
- *Perseverance.*   I will keep trying to fulfill my responsibilities regardless of obstacles. I will never give up and never quit.
- *Role model.*   I will be a consistently positive role model for my peers, colleagues, and others in my profession.
- *Mistakes.*   I will learn from my mistakes and use them as the basis for continual improvement.

Identify Your Personal Strengths and Weaknesses

Once you have established a career vision and a set of guiding principles, the next step in developing your career plan involves identifying your personal strengths and weaknesses. Your career vision provides the context for this task. The beginning point in identifying personal strengths and weaknesses is to ask yourself the following question:

> In terms of my career vision, what are my greatest strengths and weaknesses?

You answer this broad question by first answering the following more specific questions in one of two ways: "Yes" or "Needs Work." Any question answered "Needs Work" represents an area you will have to improve on because the following list of questions is based on critical factors relating to success.

1. Have you mastered the college subjects that relate most directly to your chosen career field?
2. Is personal integrity one of your best assets?
3. Are you an effective communicator?
4. Do your peers in college consider you a leader?
5. Have you identified positive role models for yourself?
6. Do you have high expectations for yourself?
7. Are you a good team builder?
8. Are you a positive change agent?
9. Do you project a winning image?
10. Are you an effective negotiator?
11. Do you apply self-discipline in a positive way?
12. Are you a good time manager?
13. Do you have a positive can-do attitude?
14. Do you know how to execute plans?
15. Are you an effective public speaker?
16. Are you a creative problem solver?
17. Are you a skilled conflict manager?
18. Are you able to live a balanced life?
19. Are you willing to help others succeed by mentoring?
20. Are you willing to persevere when work becomes difficult?

Developing Goals and Strategies

Goals translate your career vision into measurable terms. Your career-related goals should have the following characteristics:

- Broadly stated
- Specific enough to be measured
- Focused on a single issue
- Tied directly to your career vision
- Accord well with your guiding principles
- Show what you want to accomplish

SUCCESS TIP

The best way to distinguish goals from strategies is to remember this: Goals are *ends* unto themselves while strategies are *means to an end.*

Strategies are more specific than goals. They are what you have to do in order to achieve your goals. Each goal might be followed in your plan by several specific strategies. The best way to distinguish goals from strategies is to remember this: Goals are *ends* unto themselves while strategies are *means to an end.* You should have goals relating to the intermediate steps that will be necessary in order to achieve your career vision as well as goals relating to the improvement of any personal career-related weaknesses you have identified. As you make progress and as your career advances, your plan should be updated accordingly.

Sample Goals and Related Strategies

The following goals and related strategies are provided to give you examples to follow in developing your own. Remember, yours might differ from these.

Goal 1: Secure the best possible starting position in my field immediately following graduation from college.

1. Should spend three hours per week in the college's career center and online during my last year of college identifying potential employers.
2. Should develop a list of positive references I can include in my resume.
3. Should develop a comprehensive, attractive resume that is finalized prior to my last semester.
4. Should attend all career fairs sponsored by my college.
5. Should begin to interview with prospective employers no later than the beginning of my last semester in college.

Goal 2: Improve my time-management skills.

1. Should complete the next time-management seminar offered through the local chamber of commerce.
2. Should begin applying the time-management skills learned immediately while still in college.

Goal 3: Improve my public-speaking skills.

1. Should join the local chapter of Toastmasters International.
2. Should begin making oral presentations at every opportunity in my college classes to apply what I learn in Toastmasters.

Goal 4: Improve my leadership skills.

1. Should volunteer to be a group leader in college classes that require group projects.
2. Should run for a student government office in my college.

Once you have a career vision, a set of guiding principles, goals, and corresponding strategies, you have a comprehensive strategic plan for developing your career. Your plan as a college student will look vastly different from the one you should have after being in your profession for one, five, ten, and fifteen years. The key is to look on your strategic plan as a fluid document that should be changed appropriately as your circumstances change.

REVIEW QUESTIONS

1. Explain in your own words why setting high expectations for yourself is so important in building a winning career.
2. What is meant by the saying "success is a do-it-yourself project"?
3. What are the four components that should be included in your personal strategic plan for career success?
4. Write an example of a career vision for yourself.
5. Develop a set of guiding principles for yourself.
6. List two personal strengths and two weaknesses for yourself.
7. Write two career-related goals for yourself and develop at least two strategies for each goal.

DISCUSSION QUESTIONS

1. Obviously people have succeeded in their careers without developing personal strategic plans. Why, then, should you even bother with developing such a plan for yourself?
2. At this point in your life, what do you think would be realistic expectations for your career? Discuss this question with other students or friends.
3. Is there a problem with aiming too high in your career aspirations? Is it even possible to aim too high? Discuss these questions with other students or friends.

ENDNOTES

1. Louis E. Boone, *Quotable Business*, 2nd ed. (New York: Random House, Inc., 1999), 116.
2. Dennis Waitley, *The Psychology of Winning* (New York: Berkley Books, 1984), 108.
3. J. J. Fox, *How to Become a CEO* (New York: Hyperion, 1998), 7.
4. A. J. Ericsson-Jackson as quoted in *Current Biography*, Vol. 62, No. 3, March 2001, 10.
5. R. Pitino with B. Reynolds, *Success Is a Choice* (New York: Broadway Books, 1997), 45.

CHAPTER FIVE

Identify Positive Role Models and Learn From Them

It's important to pick the right role models. You're not looking for people who entertain you or people who make you feel good. You're looking for people who can help you.[1]

   Rick Pitino

A role model is someone whose good example can help you achieve your career goals. Good role models exemplify some or all of the characteristics, traits, behaviors, and abilities that promote success. Almost without exception, successful people share an important trait—they learn from the people around them. Carefully selected role models can make a significant contribution to your success by showing you in practical, observable ways what success looks like; how successful people conduct themselves, interact with others, solve problems, behave in difficult situations, and approach their work.

Consider this example. If you want to become a competitive tennis player, golfer, bowler, or volleyball player, one of the best things you can do is identify several people who play the game the way you would like to play it and observe how they do it. Then emulate their good example. This approach applies not just to people trying to establish a winning game, but also to people trying to build a winning career. By emulating the success-positive behaviors of others, you can develop your own winning ways.

HOW TO BENEFIT FROM ROLE MODELS

It has been said that the key to winning in college sports is good recruiting. This is also true with role models. The key to benefiting from role models is to select the right ones. Role models are selected because they exemplify certain characteristics, traits, behaviors, and abilities that you have identified as being important in building your winning career. Only people who can help you by their example should be chosen as role models.

Selecting good role models has been invaluable to me over the course of my career. As an athlete, I always identified coaches as well as other players who exemplified those characteristics I admired, and needed to develop in myself. When serving in the Marine Corps, I identified other Marines who had succeeded in rising rapidly up through the ranks and adopted them as role models. When working in the private sector, I selected

SUCCESS TIP

A role model is someone whose good example can help you achieve your career goals. Good role models exemplify some or all of the characteristics, traits, behaviors, and abilities that promote success. Almost without exception, successful people share an important trait—they learn from the people around them.

role models who had started their careers in entry-level jobs after college, but had risen quickly up the career ladder. When I became a college professor, I adopted as role models other college professors who were outstanding scholars and teachers. Then I emulated their best characteristics.

My all-time favorite story of a role model is Dr. James R. "Bob" Richburg, a college president I met shortly after becoming a college administrator. This individual continues to be a role model for me, especially in the area of leadership. His outstanding leadership skills began to reveal themselves in ways both large and small shortly after he became president of the college I was then serving as dean. His first official act as president was to circulate a survey instrument among all employees of the college from the janitorial staff through the top administrators in which he asked respondents to list the three best and the three worst things about the college. He then used the feedback he received to help him in formulating an improvement plan for the college.

Once Dr. Richburg and I were walking across campus with several other college administrators and he saw a piece of trash someone had casually tossed in the grass. The other administrators and I ignored the trash and walked on, but our new president walked over, picked up the trash, and deposited it in the nearest trash can. He didn't say a word to us about the trash or ask why we had not picked it up. He simply picked it up, threw it into a trash can, and rejoined our group. It was at this point that I knew this person would be a role model for me. Without saying a word, he had sent a powerful message to his administrative team: "Even the little things are important, and if a college president isn't too important to do the little things, no one is too important." Under this president, all personnel would be expected to roll up their sleeves and do what was necessary to help the college succeed. That is exactly what happened under his leadership.

SUCCESS TIP

Role models are people you can observe or study in an attempt to emulate the positive characteristics, traits, behaviors, and abilities that contribute to their success. A role model can be someone you know or it can be someone you don't know. In fact, a role model can even be a historical figure you study rather than observe.

SELECTING THE RIGHT ROLE MODELS

Mastering the courses you take in college is important. This coursework will serve as the foundation upon which to build a winning career. But just excelling in college is not enough. There are many things to learn beyond the content of your college courses if you want to build a winning career. Twenty of the most important of these are presented in this book. When selecting role models, you are looking for people who exemplify any, some, or all of these twenty behaviors and any other behaviors that will promote success. But before you begin to select role models, it is important that you understand the difference between role models and mentors—they are not the same thing.

Role models are people you can observe or study in an attempt to emulate the positive characteristics, traits, behaviors, and abilities that contribute to their success. A role model can be someone you know or it can be someone you don't know. In fact, a role model can even be a historical figure you study rather than observe. You need not have a relationship or even any contact with your role models, and you can have as many role models as you wish and change them as often as you like. Here are some people from the past I have used over the years as role models of various characteristics, traits, behaviors, and abilities in the area of leadership. For developing a vision and committing to achieving it, I have used Abraham Lincoln, who committed to saving the Union during America's Civil War. For living a life of integrity and selflessness, I have used Clara Barton, whose leadership was instrumental in the establishment of the American Red Cross. For displaying a positive can-do attitude, I have used Booker T. Washington, who transformed acres of flat, barren farmland into the most famous historically black college in the world, and who did it without government funding or assistance of any kind. Tuskegee Institute was built by the hands of Washington's students and the can-do attitude of the great man himself.

Mentors, as contrasted with role models, are people who help you actually develop the positive characteristics, traits, behaviors, and abilities you have observed in role models and wish to emulate. In other words, role models provide positive examples while mentors help you emulate those examples. Consequently, a mentor is someone with whom you have frequent contact and with whom the quality of the relationship is important. One would typically maintain a long-term relationship with a mentor rather than changing mentors frequently as can be done with role models.

What to Look for in Role Models

The selection of role models is a process that must be tailored to your specific needs. It's like going to the store to buy clothing—what fits someone else may not fit you. Consequently,

SUCCESS TIP

The selection of role models is a process that must be tailored to your specific needs. It's like going to the store to buy clothing—what fits someone else may not fit you. Consequently, it is important to select role models who are tailored to fit you.

it is important to select role models who are tailored to fit you. Here are some rules of thumb that will help you select role models that are right for you:

- Look for role models who have success-positive characteristics that you do not have or have not yet fully developed.
- Look for role models whose success can be attributed to their own merit rather than to luck or inheritance.
- Look for role models who had to overcome adversity in order to succeed—especially those who have overcome the types of adversity you have faced or might face.
- Look for role models who have successfully overcome failures or disappointments while building their careers.
- Look for role models whose abilities you can respect, not people you like. You may like a role model, but this is beside the point. The characteristics, traits, behaviors, and abilities of role models are what is important—not their likeability.

Applying What You Learn from Role Models

When you select a role model, don't try to become that person. Rather, try to emulate his success-positive characteristics, traits, behaviors, and abilities and incorporate these things into your behavior patterns. In other words, when learning from role models, don't *imitate, emulate.* For example, earlier in this chapter I explained how I have used Abraham Lincoln, Clara Barton, and Booker T. Washington as role models for various success-positive behaviors. In doing so, I never tried to become President of the United States, Director of the American Red Cross, or President of Tuskegee Institute. Rather, I tried to apply the behaviors that Lincoln, Barton, and Washington used to rise to these positions to help me rise to the various positions I have held throughout my career.

Another important principle to remember when applying what you learn from role models is this: *Your way of doing things may not be the best way.* We all have our ways of doing things, and we tend to get comfortable with those ways. However, just because we are comfortable with a certain approach does not mean it is the best approach. One of the reasons for observing and studying role models is to find better ways of doing things that are critical to building a successful career. Don't make the mistake of dropping role models from your list because the way they handle a given situation is not how you

Success Tip

Do not be deterred or put off by flaws in your role models. Role models are selected because they exhibit *certain* success-positive behaviors, not because they exhibit *all* such behaviors. Nobody is perfect. Focus on those characteristics, traits, behaviors, and abilities that are important to your success and that are demonstrated by the role model in question. Use the *menu approach* with role models. Take from the menu only what is good for you and ignore the rest.

would handle it. Rather, ask yourself if, perhaps, their approach isn't better than the one you would have taken. Remember, role models have already enjoyed a measure of success in their career or you wouldn't have selected them.

Do not be deterred or put off by flaws in your role models. Role models are selected because they exhibit *certain* success-positive behaviors, not because they exhibit *all* such behaviors. Nobody is perfect. Focus on those characteristics, traits, behaviors, and abilities that are important to your success and that are demonstrated by the role model in question. Use the *menu approach* with role models. Take from the menu only what is good for you and ignore the rest.

Remember, when choosing role models, sometimes the best example is a bad example. Sometimes you can learn as much from the mistakes of others as you can from their success-positive examples. I have identified several U.S. presidents as role models for the bad examples they exhibited in handling various aspects of their jobs. For example, I learned from these presidents not to cover up or try to hide bad news. It's better to get it out and deal with it openly and honestly than to hide it. Somehow, in the case of cover-ups, someone always seems to find out. And, when this happens, the problems only multiply.

One final way you can apply what you learn from role models is the concept of *reinforcement*. I explained earlier in this section that it is important to be open to different approaches than those you might be comfortable with. On the other hand, observing role models can also affirm that your preferred approach to a given situation is a good approach. If a role model has used an approach similar to the one you would use in a given situation and found it to be successful, you gain affirmation for your approach. This is positive reinforcement. Knowing what works is just as important as knowing what doesn't.

The story of Marsha illustrates the value of role models. When Marsha was in college, she had a definite vision for her career. Well before graduating, Marsha knew exactly where she wanted to be in her engineering career ten years after finishing college. Marsha was also honest enough with herself to realize she lacked some of the abilities necessary to achieve her career vision. She was an excellent engineering student, but needed to improve in the areas of time management, self-discipline, and conflict management. To do so, she needed role models. Unfortunately, Marsha had been unable to identify anyone who truly exemplified good time management, self-discipline, and conflict-management skills.

One day while having lunch in the Student Union of her college, Marsha mentioned to a friend her difficulty in finding good role models. Her friend asked why she was limiting search to people in the present, and explained that she found her best role

SUCCESS TIP

If a role model has used an approach similar to the one you would use in a given situation and found it to be successful, you gain affirmation for your approach. This is positive reinforcement. Knowing what works is just as important as knowing what doesn't.

models by reading the periodical *Current Biography*. Marsha began following her friend's advice from that very day onward and before long she had identified several role models who were excellent examples of good time management, self-discipline, and conflict management. As she learned about a good role model in *Current Biography*, Marsha would research the individual's life to learn even more.

Marsha continued her practice of finding role models in such sources as *Current Biography* even after graduating from college. In fact, she continued this practice throughout her career. And not only did Marsha achieve the career vision she had established while in college, but she also surpassed it.

REVIEW QUESTIONS

1. What is meant by the term *role model*?
2. Describe the type of people you should choose as role models.
3. Explain how you can benefit from having good role models.
4. Why is it so important to select the right role models?
5. Explain the differences between mentors and role models.
6. What should you look for in role models?
7. Explain the difference between *imitating* and *emulating* a role model.

SUCCESS PROFILE

Samuel B. Burkett III
Engineering Manager

Samuel B. Burkett III began his career full-time in 1975 as a database manager. He currently serves as vice president and general manager for Jacobs Sverdrup/TEAS, a company that provides engineering services to the U.S. Air Force. As the company's vice president and general manager, Burkett leads a team of 625 engineers and other personnel from a variety of different technical fields. Burkett is clearly enjoying a winning career in which he has risen to the top in his profession. But how did he do it? How did Burkett climb the career ladder from database manager to general manager?

To understand Burkett's success, one must begin with his educational preparation. Although he is now an engineering manager, Burkett did not begin college in an engineering major. That came later. First, he completed a degree in Computer Science and Statistics. This gave him a solid background in computers at a time when the world was just beginning to transform itself from analog to digital technology. Burkett then completed a master's degree—his first—in Management, which helped him advance his career in computer-related positions.

(continued)

(*continued*)

After working his way up the career ladder to the management level as a computer systems group manager, program manager, and systems analyst, Burkett, who at the time was working for an engineering company, attended college at night and completed a second master's degree—in Control and Systems Engineering. With a graduate degree in engineering, a strong computer background, a management degree, and management experience, Burkett was able to quickly move up the career ladder to the executive level.

His ongoing educational preparation has obviously been a critical enabling element in Burkett's career success. In fact, it continues to be; he is now pursuing a PhD in Engineering Management. However, in addition to formal education, Burkett has used many of the strategies set forth in this book to advance his career. In particular, he exemplifies the strategies of setting high expectations, applying good time management, and being customer-driven.

As to high expectations, Burkett knew from the outset of his career that he planned to rise to the top. He also knew that just wanting to be a corporate executive would not make him one. Hence, he developed a plan for achieving his expectations and began carrying it out. As his career advanced, Burkett—in turn—revised his plan accordingly. The plan involved completing the necessary college work and gaining appropriate types and levels of experience. It also involved being willing to *go out in order to go up*. In other words, if Burkett wanted to continue his climb up the career ladder he had to be willing to change jobs and employers when opportunities for advancement presented themselves. He did both, all while continuing his formal education at night and on weekends.

In order to pursue the course Burkett outlined for himself, it was essential to be, among other things, a good time manager. One cannot seek and accept increasing levels of responsibility at work, pursue graduate studies at night and on weekends, and balance one's family obligations successfully without being carefully attentive to good time management. Burkett exemplifies this concept, never wasting time but rather using it wisely.

Finally, Burkett is the personification of the customer-driven professional. He has to be. Most of his professional positions since beginning his career in 1975 have been customer-sensitive. This means that the work his various employers are able to secure is won competitively and retained only by satisfying the customer, which in most cases has been the U.S. military. Companies that perform services or provide products for the military operate under the strictest of guidelines, and their results are carefully monitored, measured, and evaluated. To retain contracts with the military, a company and all of its personnel must be customer-driven; they must focus intently on complete customer satisfaction and achieve it. Burkett is customer driven and, because of his example, so are the 625 people he currently leads.

DISCUSSION QUESTIONS

1. Alice is looking for one person who can be her career-development role model. Her friend Sarah does not want to limit her career-related role models to just one person. She is going to compile a list of different people who exemplify different traits she wants to emulate. Discuss these two approaches. Which do you think will be the most effective and why?

2. Jake wants to use only role models he knows personally and can talk to on a regular basis. His friend Andy tells Jake that his approach is unrealistic. They have been arguing about this issue for more than a week. Join this debate. Take one side or the other and defend your position.

3. "I want to be just like Professor Smith," says Lucy Baines. "Not me," replies Nancy Pullum. "I might emulate some of her traits and characteristics, but I don't want to be just like her." Who has the better approach to applying the lessons of role models in this situation: Lucy Baines or Nancy Pullum? Discuss your reasoning.

ENDNOTE

1. R. Pitino with B. Reynolds, *Success Is a Choice* (New York: Broadway Books, 1997), 159.

CHAPTER SIX

Be an Effective Team Player and Team Builder

Even the most talented individual cannot outperform an effective team.
   Anonymous

Teamwork is fundamental to your success in a competitive environment. The reason for this is simple and practical, as can be seen in the following quote:

Someone may be great at his or her job, maybe even the best there ever was. But what counts at work is the organization's success, not personal success. After all, if your organization fails, it does not matter how great you were; you are just as unemployed as everyone else.[1]

WHAT IS A TEAM?

A team is a group of people with a common, collective goal. The collective goal aspect of teams is critical. This point is evident in the performance of athletic teams. For example, a basketball team in which one player hogs the ball, behaves like a prima donna, and pursues his own personal goals (high point total, MVP status, publicity, or something else) will rarely win against a team of players, all of whom pull together toward the collective goal of winning the game. Before you can be a team builder you must first be a team player. People who are able to make teams win are more likely to become winners themselves.

An individual I'll call Andrea knew about being a team player and a team builder. While in high school, she had been a member of a championship volleyball team. She and her teammates were shorter than the players on all of the other teams they played. With no standout players and with no one to provide height at the net, Andrea and her teammates had to compensate by working well together as a team.

 ## SUCCESS TIP

> Before you can be a team builder you must first be a team player. People who are able to make teams win are more likely to become winners themselves.

Andrea and her teammates quickly learned that effective teamwork could compensate for their lack of height and speed. Consequently, they emphasized teamwork while striving to continually improve through team building. Andrea found she was actually better at team building than volleyball. She was an excellent motivator, a good leader, and a selfless player who always put the team first and helped her teammates play better. Consequently, on and off the floor, Andrea became an extension of the coach, and was eventually elected captain by her teammates as a result.

Andrea did not play volleyball in college, but she did put her team-related skills to work in the classroom. When group activities were assigned, Andrea quickly emerged as a leader who worked hard to motivate her classmates to perform at levels higher than they thought themselves capable of. Consequently, no one who knew Andrea was surprised at her rapid ascent up the career ladder following graduation from college.

In the organization that employed her, Andrea did what she had been doing since high school. She quickly established herself as a team player, team builder, and team leader. Any team Andrea served on soon gained a reputation for getting the job done right and on time. Andrea's team-related skills ensured that any team she served on would perform better than it would have without her. This, in turn, helped propel Andrea steadily up the career ladder until, finally, the team she captained was her own company.

BUILDING TEAMS AND MAKING THEM WORK EFFECTIVELY

Part of building a successful team is choosing team members wisely. This section describes strategies you can apply when selecting team members, assigning responsibilities, creating a mission statement, and developing collegial relations among team members. Remember, work teams are like sports teams in that it does matter who is on them.

Choosing Team Members

When putting together a team, the first step is to identify all potential team members. This is important because there will often be more potential team members than the number of members actually needed. After the list has been compiled, team members can be selected. However, care should be taken to ensure a broad mix. This rule should be adhered to even if there are no volunteers, and team members must be drafted. The more likely case is that there will be more volunteers than openings on most teams.

A reality of the workplace is that you won't always be able to choose your team members. Sometimes they are chosen for you. In either case, what follows applies.

Responsibilities of Team Leaders

Becoming a team leader is often the first step you will take up the career ladder. Consequently, when you get an opportunity to lead a team it is important to get it right. Scholtes lists the following as responsibilities of team leaders:[2]

■ Serve as the official contact between the team and the rest of the organization.

■ Serve as the official record keeper for the team. Records include minutes, correspondence, agendas, and reports. Typically, the team leader will appoint a recorder to take

SUCCESS TIP

After your team has been formed, the next step is to draft its mission statement. This is a critical step in the life of a team. The mission statement explains the team's reason for being. A mission statement is written in terms that are broad enough to encompass everything the team will be expected to do, but specific enough that progress can be measured.

minutes during meetings. However, the team leader is still responsible for distributing and filing minutes and making sure they are accurate.

- Serve as a full-fledged team member, but take care to avoid dominating team discussions.
- Implement team recommendations that fall within the team leader's realm of authority, and work with upper management to implement those that fall outside of it.

Creating the Team's Mission Statement

After your team has been formed, the next step is to draft its mission statement. This is a critical step in the life of a team. The mission statement explains the team's reason for being. A mission statement is written in terms that are broad enough to encompass everything the team will be expected to do, but specific enough that progress can be measured. The following sample mission statement meets both of these criteria:

> The purpose of this team is to reduce the time between when an order is taken and when it is filled while simultaneously improving the quality of products shipped.

This statement is broad enough to encompass a wide range of activities and to give team members plenty of room in which to operate. The statement does not specify by how much throughput time will be reduced or how much quality will be improved. The level of specificity comes in the goals set by the team (e.g., reduce engineering time by fifteen percent within six months; improve the customer satisfaction rate to one hundred percent within six months). Goals follow the mission statement and explain it more fully in quantifiable terms.

This sample mission statement is written in broad terms, but is specific enough that team members know they are expected to simultaneously improve both productivity and

SUCCESS TIP

Competence, trust, communication, and mutual support form the foundation on which effective teamwork is built. Any time devoted to improving these factors is a good investment on your part.

SUCCESS TIP

Team leaders who begin team-building activities without first assessing strengths and weaknesses run the risk of wasting resources in an attempt to strengthen characteristics that are already strong, while at the same time overlooking characteristics that are weak.

quality. It also meets one other important criterion: simplicity. Any employee could understand this mission statement. It is brief and to the point, but comprehensive.

Team leaders should keep these criteria in mind when developing mission statements: balance broadness, specificity, and simplicity. A good mission statement is a tool for communicating the team's purpose, both within the team and throughout the organization.

Developing Collegial Relationships

A team works most effectively when individual team members form positive, mutually supportive peer relationships. These are *collegial relationships,* and they can be the difference between a high-performance team and a mediocre one. Learning to develop collegial relationships in teams will make you a more valuable member, which in turn will promote success—both yours and the team's. Scholtes recommends the following strategies for building collegial relationships among team members:[3]

■ Help team members understand the importance of honesty, reliability, and trustworthiness. Team members must trust each other and know that they can count on each other.
■ Help team members develop mutual confidence in their work ability.
■ Help team members understand the pressures to which other team members are subjected. It is important for team members to be supportive of peers as they deal with the stresses of the job.

These are the basics. Competence, trust, communication, and mutual support form the foundation on which effective teamwork is built. Any time devoted to improving these factors is a good investment on your part.

FOUR-STEP APPROACH TO TEAM BUILDING

Effective team building is a four-step process:

1. Assess
2. Plan
3. Execute
4. Evaluate

To be a little more specific, the team-building process proceeds along the following lines: (a) assess the team's developmental needs (e.g., its strengths and weaknesses), (b) plan team-building activities based on the needs identified, (c) execute the planned team-building activities, and (d) evaluate results. These steps are explained further in the following sections.

Assessing Team Needs

If you were the new coach of a baseball team about which you knew very little, what is the first thing you would want to do? Most coaches in such situations would begin by assessing the abilities of their team members. What are our weaknesses? What are our strengths? With these questions answered, the coach will know how best to proceed with team-building activities.

This same approach can be used in the workplace. A mistake often made by team leaders is beginning team-building activities without first assessing the team's development needs. Resources are often limited in organizations. Consequently, it is important to use them as efficiently and effectively as possible. Team leaders who begin team-building activities without first assessing strengths and weaknesses run the risk of wasting resources in an attempt to strengthen characteristics that are already strong, while at the same time overlooking characteristics that are weak. Understanding the importance of assessing team needs will give you an advantage over other team leaders.

For workplace teams to be successful, they should have at least the following characteristics:

- Clear direction that is understood by all members (e.g., mission and goals).
- *Team players* on the team (e.g., team first—me second).
- Fully understood and accepted accountability measures (e.g., evaluation of performance).

Planning Team-Building Activities

Team-building activities should be planned based on the results of your assessment of needs. For example, say the team appears to be floundering because it lacks direction. Clearly, part of the process of building this team must be either developing a mission statement or explaining the existing statement. If team members don't seem to be as interested in their performance as they should be, accountability measures might be weak. Regardless of what your assessment reveals about the team, use that information to plan team-building activities that will enhance the team's performance.

Executing Team-Building Activities

Team-building activities should be implemented on a just-in-time basis. This means it is best to provide team-building activities only after an actual team has been formed. Training everyone in an organization in case they are put on a team later is not an effective approach. Like any kind of training, teamwork training will be forgotten unless it is put to use immediately. Consequently, the best time to provide teamwork training is after

SUCCESS TIP

If employees are going to be expected to work together as a team, leaders have to realize that teams are not bossed—they are coached.

a team has been formed and given its charter. In this way, team members will have opportunities to immediately apply what they are learning.

Team building is an ongoing process. The idea is to make a team better and better as time goes by. Consequently, basic teamwork training is provided as soon as a team is chartered. All subsequent team-building activities should be based on the results of a needs assessment.

Evaluating Team-Building Activities

If team-building activities have been effective, weak areas pointed out by the needs assessment should have been strengthened. A simple way to evaluate the effectiveness of team-building activities is to wait a sufficient amount of time to allow the team's performance to improve and then conduct the needs assessment again.

If this evaluation shows that sufficient progress has been made, nothing more is required. If not, additional team-building activities are needed. If a given team-building activity appears to have been ineffective, get the team together and discuss it. Use the feedback from team members to identify weaknesses and problems and use the information to ensure that team-building activities become more effective. Involving the team in evaluating team-building activities is itself a team-building activity.

TEAMS AREN'T BOSSED—THEY ARE COACHED

If employees are going to be expected to work together as a team, leaders have to realize that teams are not bossed—they are coached. Team leaders need to understand the difference between bossing and coaching. Bossing, in the traditional sense, involves giving orders and evaluating performance. Bosses approach the job from an "I'm in charge—do what I say" perspective.

Coaches, on the other hand, are facilitators of team performance, team development, and team improvement. They lead the team in such a way that it achieves peak performance levels on a consistent basis. You can be an effective coach of work teams by doing the following:

- Give your team a clearly defined charter (mission and goals).
- Make team development and team building a continual activity.
- Mentor team members.
- Promote mutual respect between you and team members and among team members.

- Work to make human diversity within a team a plus.
- Handle conflicts in teams.

Clearly Defined Charter (Mission and Goals)

You can imagine a basketball, soccer, or track coach calling her team together and saying, "This year we have one overriding purpose—to win the championship." In one simple statement this coach has clearly and succinctly defined the team's charter. All team members now know that everything they do this season should be directed toward winning the championship. Coaches of work teams should be just as specific in explaining the team's mission to their team members.

Team Development/Team Building

Work teams should be similar to athletic teams when it comes to team development and building. Regardless of the sport, athletic teams practice constantly. During practice, coaches work on developing the skills of individual team members and the team as a whole. Team development and team-building activities are a constant and they should go on forever. A team should never stop getting better. Coaches of work teams should follow the lead of their athletic counterparts. Developing the skills of individual team members and building the team as a whole should be a normal part of the job—a part that takes place regularly and never stops.

Mentoring

Good coaches are mentors. This means they establish a helping, caring, nurturing relationship with team members. Developing the capabilities of team members, improving the contribution individuals make to the team, and helping team members advance their careers are all mentoring activities. Effective mentors help team members in the following ways:[4]

- Developing their job-related competence
- Building character
- Teaching them the corporate culture
- Teaching them how to get things done in the organization
- Helping them understand other people and their viewpoints

SUCCESS TIP

Good coaches are mentors. This means they establish a helping, caring, nurturing relationship with team members. Developing the capabilities of team members, improving the contribution individuals make to the team, and helping team members advance their careers are all mentoring activities.

- Teaching them how to behave in unfamiliar settings or circumstances
- Giving them insight into differences among people
- Helping them develop success-oriented values

Mutual Respect

It is important for team members to respect their coach, for the coach to respect his team members, and for team members to respect each other. According to Robert H. Rosen, "Respect is composed of a number of elements, that like a chemical mixture, interact and bond together."[5]

- *Trust made tangible.*   Trust is built by (a) setting the example, (b) sharing information, (c) explaining personal motives, (d) avoiding both personal criticisms and personal favors, (e) handing out sincere rewards and recognition, and (f) being consistent in disciplining.
- *Appreciation of people as assets.*   Appreciation for people is shown by (a) respecting their thoughts, feelings, values, and fears, (b) respecting their desire to lead and follow, (c) respecting their individual strengths and differences, (d) respecting their desire to be involved and to participate, (e) respecting their need to be winners, (f) respecting their need to learn, grow, and develop, (g) respecting their need for a safe and healthy workplace that is conducive to peak performance, and (h) respecting their personal and family lives.
- *Communication that is clear and candid.*   Communication can be made clear and candid if coaches will do the following: (a) open their eyes and ears—observe and listen, (b) say what you want and say what you mean (be tactfully candid), (c) give feedback constantly and encourage team members to follow suit, and (d) face conflict within the team head-on; that is, don't let resentment among team members simmer until it boils over—handle it now.
- *Ethics that are unequivocal.*   Ethics can be made unequivocal by (a) working with the team to develop a code of ethics, (b) identifying ethical conflicts or potential conflicts as early as possible, (c) rewarding ethical behavior, (d) disciplining unethical behavior, and doing so consistently, and (e) when bringing in new team members, making them aware of the team's code of ethics. In addition to these strategies, the coach should set a consistent example of unequivocal ethical behavior.

Human Diversity

Human diversity can be a plus. Sports and the military have typically led American society in the drive for diversity, and both have benefited immensely as a result. To list the contributions to either sports or the military made by people of different genders, races, religions, etc. would be a gargantuan task. Fortunately, leading organizations in the United States are following the positive example set by sports and the military. The smart ones have learned that most of the growth in the workplace will be among women, minorities, and immigrants. These people will bring new ideas and varying perspectives, precisely what an organization needs to stay on the razor's edge of competitiveness. However, in spite of

SUCCESS TIP

Conflict will occur in even the best of teams. Even when all team members agree on a goal, they can still disagree on how best to accomplish it. Few things will further your career more than being able to resolve conflict in a positive way.

steps already taken toward making the American workplace both diverse and harmonious, wise coaches understand that people—consciously and unconsciously—sometimes erect barriers between themselves and people who are different from them. This tendency can quickly undermine the trust and cohesiveness on which teamwork is built. To keep this from happening on teams you lead, try the following strategies:

- *Identify the specific needs of different groups.* Ask women, ethnic minorities, and older workers to describe the unique inhibitors they face. Make sure all team members understand these barriers, and then work together as a team to eliminate, overcome, or accommodate them.

- *Confront cultural clashes.* Meet conflict among team members head-on and immediately. This approach is particularly important when the conflict is based on diversity issues. Conflicts that grow out of religious, cultural, ethnic, age, and/or gender-related issues are more potentially volatile than everyday disagreements over work-related concerns. Consequently, conflict that is based on or aggravated by human differences should be confronted promptly. Few things will polarize a team faster than diversity-related disagreements that are allowed to fester and grow.

- *Eliminate institutionalized bias.* Consider the following example. A company workforce that had historically been predominantly male now has turned into a workforce in which women are the majority. However, the facility still has ten rest rooms for men and only two for women. This imbalance is an example of institutionalized bias. Teams may find themselves unintentionally slighting members, simply out of habit or tradition. This is the concept of *discrimination by inertia.* It happens when the demographics of a team change but its habits, traditions, procedures, and work environment do not.

An effective way to eliminate institutional bias is to circulate a blank notebook and ask team members to record—without attribution—instances and examples of institutional bias. After the initial circulation, repeat the process periodically. You can use the input collected to help eliminate institutionalized bias. By collecting input directly from team members and acting on it promptly, you can ensure that discrimination by inertia is not creating or perpetuating debilitating resentment.

Handling Conflict in Teams

Conflict will occur in even the best of teams. Even when all team members agree on a goal, they can still disagree on how best to accomplish it. Few things will further your career more than being able to resolve conflict in a positive way. Lucas recommends the following strategies for preventing and resolving team conflict:[6]

- Plan and work to establish a culture where individuality and dissent are in balance with teamwork and cooperation.
- Establish clear criteria for deciding when decisions will be made by individuals and when they will be made by teams.
- Don't allow individuals to build personal empires or to use the organization to advance personal agendas.
- Encourage and recognize individual risk-taking behavior that breaks the organization out of unhelpful habits and negative mental frameworks.
- Encourage healthy, productive competition, and discourage unhealthy, counterproductive competition.
- Recognize how difficult it can be to ensure effective cooperation, and spend the energy necessary to get just the right amount of it.
- Value constructive dissent, and encourage it.
- Assign people of widely differing perspectives to every team or problem.
- Reward and recognize both dissent and teamwork when they solve problems.
- Reevaluate the project, problem, or idea when no dissent or doubt is expressed.
- Avoid hiring people who think they don't need help or who don't value cooperation.
- Ingrain into new employees the need for balance between the concepts of *cooperation* and *constructive dissent*.
- Provide ways for employees to say what no one wants to hear.
- Realistically and regularly assess the ability and willingness of employees to cooperate effectively.
- Understand that some employees are going to clash, so determine where this is happening and remix rather than waste precious organizational energy trying to get people to like each other.
- Ensure that the organization's value system and reward/recognition systems are geared toward cooperation with constructive dissent rather than dog-eat-dog competition or cooperating at all costs.
- Teach employees how to manage (not let it get out of hand) both dissent and agreement.
- Quickly assess whether conflict is healthy or destructive, and take immediate steps to encourage the former and resolve or eliminate the latter.

REVIEW QUESTIONS

1. In your own words, define the term *team*.
2. Explain how to choose team members when forming a work team.
3. List and explain the main responsibilities of team leaders.
4. Write an example of a mission statement for a hypothetical team.
5. Explain the four-step approach to team building.

6. What is the difference between a *boss* and a *coach* in the language of teamwork?
7. List the ways effective mentors help their team members.
8. How can team leaders build trust in their teams?
9. Explain how team leaders can ensure that human diversity is a *plus* in their teams' performance.
10. List at least five strategies for preventing and resolving conflict in teams.

DISCUSSION QUESTIONS

1. "My team members don't seem to get along," said Faye Sims to her fellow supervisor, Dot Moore. "Do you have any suggestions that will help me create a more collegial environment in my team?" Join in this discussion. Discuss various strategies Sims might use to encourage collegiality in her team.
2. While driving together to a meeting one day, Al Carver's boss told him, "Al, you would get a lot more out of your team if you would stop bossing and start coaching." Carver wasn't sure what his boss meant, and he didn't want to appear ignorant, so he didn't ask. Discuss what you think Carver's boss meant.
3. Tina Smith is growing increasingly frustrated with her team. There is a lot of petty bickering and discord among the members. She thought diversity was supposed to be a good thing in teams. So far it has done nothing but create problems for her. Discuss how Tina Smith might build cohesiveness among her team members and eliminate the petty bickering that is now occurring.
4. Pablo Rodriguez has just been promoted to team leader and given the assignment of eliminating the counterproductive conflict in his team. Apparently his predecessor was unable to manage conflict among team members; conflict that had gotten out of hand and was undermining the team's productivity. Discuss some ways Rodriguez can resolve the conflict in his team and prevent it from resurfacing as a problem once it has been resolved.

ENDNOTES

1. Perry L. Johnson, Rob Kantner, and Marcia A. Kikora, *TQM Team-Building and Problem-Solving* (Southfield, MI: Perry Johnson, Inc., 1990), 1-1.
2. Peter R. Scholtes, *The Team Handbook* (Madison, WI: Joiner Associates, 1992), 3–10.
3. Ibid., 3–8.
4. Gordon F. Shea, *Mentoring* (New York: AMACOM, American Management Association, 1994), 49–59.
5. Robert H. Rosen, *The Healthy Company* (New York: Perigee Books, Putnam, 1991), 24.
6. James R. Lucas, *Fatal Illusions* (New York: AMACOM, American Management Association, 1996), 160–161.

CHAPTER SEVEN

Be a Positive Change Agent

To bring about change a leader must be willing to swim against the current of custom, norms, and culture.

Anonymous

Few things call so clearly for effective leadership as organizational change. This is because it takes an effective leader to overcome an inherent characteristic I call *organizational inertia.* You can use this fact to advance your career if you become a positive change agent— a person who can facilitate change and continual improvement in an organization.

Organizations are like the "body at rest" in the scientific principle of inertia. An organization with a static culture (a body at rest) will tend to remain static until sufficient leadership is applied to get it moving. Another way to state this principle as it applies to organizations is, *An organization's culture, once established, will tend to perpetuate itself until sufficient leadership is applied to change it.* An excellent way to advance your career is to learn how to provide leadership to ensure that organizations from the smallest teams to the largest corporations not only change to stay current, but actually get out in front of change.

IMPORTANCE OF LEADERSHIP IN CHANGE

In a competitive and rapidly changing marketplace, organizations are constantly looking for ways to keep up, stay ahead, and/or set new directions. What can you do to play a positive role in the process? First, you will have to advance sufficiently to be in a leadership position (e.g., team leader, supervisor, or manager). Then, the following strategies will help you to be a positive change agent:

- Develop an inviting change picture (how things will be after the change).
- Take responsibility for change.
- Communicate your change picture.
- Identify and eliminate barriers to the change.
- Establish incremental progress points and monitor progress constantly.

Carla was accustomed to change. As the oldest sibling in a military family, she had been uprooted from her comfortable surroundings many times as her father's Air Force career took the family to a succession of different assignments around the globe. In fact, it seemed to Carla that—during her childhood years—just when she would get comfortable with a new home, school, and friends, her dad would be transferred and Carla would have to start all over again. Not only did she have to adjust herself to a succession of new homes, schools, and friends, but as the oldest sibling in the family, Carla also had to help her brother and sister adjust. Over time, she got good at it.

Carla was especially effective at "painting" in words an attractive change picture for her siblings. Before each family relocation, Carla would research the new community to identify where her brother and sister would attend school and to locate interesting attractions in the new community. Then she would use her research findings to develop an appealing vision of how things would be after the move. Then Carla gave each of her siblings a list of tasks to accomplish immediately following the move—tasks she knew would make the move less difficult for them. The list always included such assignments as getting their rooms organized and decorated within just two days of moving in, visiting their new school to get familiar with the campus, making at least one new friend within a week of arriving at their new home, and identifying one thing they liked about the new neighborhood within a week of moving in. She also insisted that her siblings share their findings with the family.

After graduating from college, Carla went to work as an engineer. Within a year, she was a team leader. Shortly after Carla was named a team leader, her company was acquired by a competitor. The acquisition was disruptive, and many employees of Carla's company experienced great difficulty as they struggled to cope with the transition. But for Carla, the change was just business as usual. In helping her team members make the transition, she applied the same strategies she had always used with her siblings following a family move. Carla's team performed so well during and after the transition that Carla earned a promotion under the new management—the first of many. Being a positive change agent and effective change leader when her organization underwent a major change helped propel Carla past her contemporaries on the career ladder.

FACILITATING CHANGE AS A LEADERSHIP FUNCTION

The following statement by management consultant Donna Deeprose carries a particularly relevant message for people who want to be positive change agents:

> In an age of rapidly accelerating technology, restructuring, repositioning, downsizing, and corporate takeovers, change may be the only constant. Is there anything you can do about it? Of course there is. You can make change happen, let it happen to you, or stand by while it goes on around you.[1]

Deeprose divides people into three categories, based on how they handle change: driver, rider, or spoiler.[2] People who are drivers lead their organizations in new directions in response to change. People who just go along, reacting to change as it happens rather then getting out in front of it, are riders. People who actively resist change are spoilers. Do you want to be a driver? Deeprose gives examples of how a driver would behave in a variety of situations:[3]

SUCCESS TIP

> To survive and thrive in a competitive environment, organizations must be able to effectively anticipate and respond to change. However, the most successful organizations don't just respond to change—they get out in front of it. Helping your organization get out in front of change will help you get out in front of the competition for promotions and raises.

- In viewing the change taking place in an organization, drivers stay mentally prepared to take advantage of the change.
- When facing change about which they have misgivings, drivers step back and examine their own motivations.
- When a higher manager has an idea that has been tried before and failed, drivers let the boss know what difficulties were experienced earlier and offer suggestions for avoiding the problems this time.
- When a company announces major changes in direction, drivers find out all they can about the new plans, communicate what they learn to employees, and solicit input to determine how to make a contribution to the achievement of the company's new goals.
- When a change will affect other departments, drivers go to these other departments and explain the change in their terms, solicit their input, and involve them in the implementation process.
- When demand for their unit's work declines, drivers solicit input from users and employees as to what modifications and new products or services might be needed and include the input in a plan for updating and changing direction.
- When an employee suggests a good idea for change, drivers support the change by justifying it to higher management and using their influence to obtain resources for it while countering opposition to it.
- When their unit is assigned a new, unfamiliar task, drivers delegate the new responsibilities to their team members and make sure they get the support and training needed to succeed.

These examples show that a driver is a person who exhibits the leadership characteristics necessary to play a positive, facilitating role in helping employees and organizations successfully adapt to change on a continual basis. Being a driver is good for your career.

HOW TO LEAD CHANGE

A critical responsibility in globally competitive organizations is leading change. The ability to lead change will do much to advance your career. To survive and thrive in a competitive environment, organizations must be able to effectively anticipate and respond to change. However, the most successful organizations don't just respond to change—they get out in front of it. Helping your organization get out in front of change will help you get out in front of the competition for promotions and raises.

Change is a constant in today's global business environment. Consequently, organizations must structure themselves for change. In other words, organizations must *institutionalize* the process of change. The following strategies taken together comprise an effective model for leading change.[4] You can stand out from the crowd in a positive way by helping your organization implement this model.

Understand the Reality of Continual Change

Change is not something organizations do because they want to or because they get bored with the status quo. Rather, it is something they do because they must. Every organization that operates in a competitive environment is forced by macroeconomic conditions to constantly reduce costs, improve quality, enhance product attributes, increase productivity, and identify new markets. None of these things can be accomplished without continually improving (changing) the way things are currently done.

Since change is the most constant element on the radar screen of the modern organization, why do so many miss the point? A typical reaction to this question is that human beings don't like change. But in reality this is not always the case. People who object to change often object to how it is handled, not the fact that it's happening. Kotter identifies the following reasons that people in an organization may not understand the need for change:[5]

- Absence of a major crisis
- Low overall performance standards
- No view of the big picture
- Internal evaluation measures that focus on the wrong benchmarks
- Insufficient external feedback
- A "kill the messenger" mentality among managers
- Over-focus among employees on the day-to-day stresses of the job
- Too much "happy talk" from executives

An organization's senior leadership team is responsible for dealing with these factors. However, until you are an executive, you can play an important and helpful role by being a facilitator at your level.

Performance standards should be based on what it takes to succeed in the global marketplace. Every organization should have a strategic plan that explains the "big picture," and every employee should know the big picture. You can help by making sure your teammates and direct reports understand their roles in the big picture. Internal evaluation measures should mirror overall performance standards in that they should ensure globally competitive performance.

Organizations cannot succeed in a competitive environment without systematically collecting, analyzing, and using external feedback. This is how an organization knows what is going on. External feedback is the most effective way to identify the need for

SUCCESS TIP

Although a visible and visionary individual can certainly be a catalyst for organizational change, the reality is that one person rarely changes an organization.

change. Every employee in an organization is a potential agent for change. Employees attend conferences, read professional journals, participate in seminars, browse the Internet, and talk to colleagues. You can help by being sensitive to cues in your professional environment that indicate the need for change, and by sharing your information with your organization's management team.

As a rule, employees will focus most of their attention on their day-to-day duties, which is how it should be. Consequently, you should make a special effort to communicate with your colleagues and direct reports about market trends and other big-picture issues. All such communications should be accurate, thorough, and honest. Don't go overboard and create panic, but try to keep colleagues, teammates, and direct reports focused on the need for continual improvement (change).

Establish and Charter the Steering Committee

When organizations are about to go through a major change, a wise strategy is to establish a steering committee with responsibility for making the change happen.

> Major transformations are often associated with one highly visible leader. Consider Chrysler's comeback from near bankruptcy . . . and we think of Lee Iacocca. Mention Wal-Mart's ascension from small fry to industry leader, and Sam Walton comes to mind. Read about IBM's efforts to renew itself and the story centers on Lou Gerstner. After a while, one might easily conclude that the kind of leadership that is so critical to any change could come only from a single larger-than-life person.[6]

Although a visible and visionary individual can certainly be a catalyst for organizational change, the reality is that one person rarely changes an organization. The media like to create the image of the knight in shining armor that single-handedly saves the company. This story makes good press, but it rarely squares with reality.

SUCCESS TIP

Reading engineering journals, attending conferences, studying global markets, and even reading the newspaper can help you identify trends that might affect your organization. So can staying in touch with your organization's customers.

Organizations that do the best job of handling change are those that establish a steering committee. The steering committee is a team of people who are committed to the change in question and who can make it happen. Every member of the team should have one or more of the following characteristics:

- *Authority.*   Members should have the authority necessary to make decisions and commit resources.
- *Expertise.*   Members should have expertise that is pertinent in terms of the change in question so that informed decisions can be made.
- *Credibility.*   Members must be well respected by all stakeholders so they will be listened to and taken seriously.
- *Leadership.*   Members should have the leadership qualities necessary to drive the effort. These qualities include those listed here plus influence, vision, commitment, perseverance, and persuasiveness.

If your organization establishes a steering committee to oversee a major change, try to secure a position as a member of it. If you cannot be a member of the steering committee, stay informed of its activities and do all you can to help it succeed.

Establish Antenna Mechanisms

Leading change is about driving change rather than letting it drive you. To do this, organizations must have mechanisms for sensing trends that will generate future change. You can establish an antenna mechanism for your organization. These "antenna" mechanisms can take many forms, and the more the better. Reading engineering journals, attending conferences, studying global markets, and even reading the newspaper can help you identify trends that might affect your organization, and help you stay in touch with your organization's customers.

For example, a computer company that markets to colleges and universities learned that two large institutions had adopted a plan to stop purchasing personal computers. Instead, they intended to require all students to purchase their own laptops. This potential threat to the computer company was identified by one of its marketing representatives while making a routine call on these universities.

This advance information allowed the computer company to quickly develop and implement a plan for getting out in front of what might eventually become a nationwide trend. The company now markets laptops directly to students through bookstores at both universities and is promoting the idea to colleges and universities nationwide.

You should have your antennae tuned to the world outside and bring anything of interest to the attention of your organization's steering committee.

Develop a Vision (Change Picture)

People want to know how things will be. How will things look after the change in question has been made? The organization's vision answers this question. Kotter calls a vision a "sensible and appealing picture of the future."[7] The following scenario illustrates how having a cogent vision for change can help people in an organization accept the change more readily.

Two tour groups are taking a trip together. Each group has its own tour bus and group leader. The route to the destination and all stops along the way have been meticulously planned for maximum value, interest, and enjoyment. All members of both groups have agreed to the itinerary and are looking forward to every stop. Unfortunately, several miles before the first scheduled attraction, the tour group leaders receive word that a chemical spill on the main highway has caused a detour that will, in turn, require changes in the itinerary.

The tour group leader for Group A simply acknowledges the message and tells the bus driver to take the alternate route. To the members of Group A he says only that an unexpected detour has forced a change of plans. With no more information than this to go on, the members of Group A are confused and quickly become unhappy.

The tour group leader for Group B, however, is a different sort. She acknowledges the message and asks the driver to pull over. She then tells the members of Group B the following:

> Folks, we've had a change of plans. A chemical spill on the highway up ahead has closed down the route on which most of today's attractions are located. Fortunately, this area is full of wonderful attractions. Why don't we just have lunch on me at a rustic country restaurant near here? While you folks are enjoying lunch, I'll hand out a list of new attractions I know you'll want to see. We aren't going to let a little detour spoil our fun!

Because they could see how things would look after the change, the members of Group B accepted it and were satisfied. However, the members of Group A, because they were not fully informed, became increasingly frustrated and angry. As a result, they went along with the changes only reluctantly and, in several cases, begrudgingly. Group A's tour leader kept his clients in the dark. Group B's leader gave hers a vision. To advance your career, follow the example of the tour leader for Group B. When a major change must be made, give your team members a vision—paint a compelling change picture for them. The five characteristics of an effective vision for change are as follows:

1. *Imaginable.* It conveys a picture that others can see of how things will be after the change.
2. *Desirable.* It points to a better tomorrow for all involved.
3. *Feasible.* It is realistic and attainable.
4. *Flexible.* It is stated in terms that are general enough to allow for initiative in responding to ever-changing conditions.
5. *Communicable.* It can be explained to an outsider who has no knowledge of the business.

Until you are an executive, you won't develop change visions, but regardless of your level in an organization, make sure you understand the vision.

Communicate the Vision to All Involved (Stakeholders)

People cannot buy into the change vision unless they know about it. Therefore, the vision must be communicated to all stakeholders. A good communication approach will have at least the following characteristics:

- Simplicity
- Repetitiveness

- Multiple formats
- Feedback mechanisms

The simpler the message, the better. Be direct and get to the point. Don't beat around the bush or attempt any type of linguistic subterfuge—what politicians and journalists call *spin*. Whatever your level in the organization, you can help communicate the change vision to others in the organization.

Repetition is critical when communicating a new message. Messages are like flies. If a fly buzzes past your face just once and moves on, you will probably take little notice of it. But if it keeps coming back persistently and refuses to go away, before long you will take notice of it. Like the fly in this example, in order to be noticed, a message must be repeated.

It is also important to use multiple communication formats, such as small-group meetings, large-group meetings, newsletter articles, fliers, bulletin board notices, video-taped messages, e-mail, and a variety of other methods. Different people prefer different formats. Consequently, a combination of visual, reading, and listening vehicles will typically be the most effective.

Regardless of the communication formats used, one or more feedback mechanisms should be put in place. In face-to-face meetings, the feedback can be spontaneous and direct. This is always the best form of feedback. However, telephone, facsimile, e-mail, and written feedback can also be valuable. When you are trying to communicate a change message, remember these factors: simplicity, repetitiveness, multiple formats, and feedback mechanisms.

Incorporate the Change Process

Once an organization has gone through a major change, both the change itself and the process of change should be incorporated as part of the organization's culture. You can play an important role in making this happen at your level. In other words, two things need to happen. First, the change that has just occurred should become the normal way of doing business. Second, the process of change explained in this section should be institutionalized. You will want to play a leadership role in the institutionalization of change once you are in a leadership position.

Anchoring the new change in the organizational culture is critical. If this does not occur, the organization will quickly begin to backslide and retrench. Remember this. Following up to make sure the change "sticks" will set you apart from the competition for career advancement. The following strategies will help you anchor a major change in the organization's culture:

- *Showcase the results.*   In the first place, a change must work in order to be accepted. The projected benefits of making the change should be showcased as soon as they are realized and, of course, the sooner the better.
- *Communicate constantly.*   Do not assume that stakeholders will automatically see, understand, and appreciate the results gained by making the change. Talk constantly about results and their corresponding benefits.

SUCCESS TIP

Antenna mechanisms continue to anticipate change all the time, forever. They feed what they see into the model, and the organization works its way through each step explained here. The better an organization becomes at doing this the more successful it will be at competing in the global marketplace. The better you become at helping your organization make changes, the farther you will go in your career.

■ *Remove resistant employees.* If employees are still fighting the change after it has been made and is producing good results, give them the "get with it or get out" option. This approach might seem harsh, but employees at all levels are paid to move an organization forward, not hold it back.

Institutionalizing the process of change is an important and final element of this step. Change is not something that happens once and then goes away. It is a constant in the lives of every person, in every organization. Consequently, the change facilitation model explained here should become part of normal business operations, and you should develop a reputation for doing your part to implement the model. Antenna mechanisms continue to anticipate change all the time, forever. They feed what they see into the model, and the organization works its way through each step explained here. The better an organization becomes at doing this the more successful it will be at competing in the global marketplace. The better you become at helping your organization make changes, the farther you will go in your career.

REVIEW QUESTIONS

1. Explain how the scientific principle of inertia applies to organizations facing change.
2. Explain several strategies that will help you to be a positive change agent in an organization.
3. Use your own words to describe the concept of the *driver* as it relates to change in an organization.
4. What are some reasons that people in an organization might not understand the need for change?
5. What are the ideal characteristics of people who would serve on an organization's steering committee for change?
6. Explain the concept of the *change picture.*
7. What are the characteristics of good communication as it relates to change management?
8. When trying to anchor a change that has been made in an organization's culture, what strategies should you use?

DISCUSSION QUESTIONS

1. "No organization ever changes without good leadership." Debate and discuss this contention. Is it correct or incorrect? Explain your reasoning.
2. Assume you are an engineering supervisor at XYZ Company and the CEO just announced plans for a merger with ABC. You are known as a *driver* of change. Discuss what you should do to alleviate the concerns of the employees in your team.
3. You have just become the CEO of a small engineering company that needs to make major changes. Discuss the challenge ahead of you and how you propose to handle it.
4. Assume your college or university plans to make major changes that will affect students and faculty members. If you were the institution's president, how would you go about communicating your vision for change to students and faculty members? Discuss strategies with fellow students or friends.

ENDNOTES

1. Donna Deeprose, "Change: Are You a Driver, a Rider, or a Spoiler?" *Supervisory Management,* February, 1990, 3.
2. Ibid.
3. Ibid.
4. John P. Kotter, *Leading Change* (Boston: Harvard Business School Press, 1996), 3.
5. Ibid., 40.
6. Ibid., 51.
7. Ibid., 71.

CHAPTER EIGHT

Project a Winning Image

Some succeed by what they know; some by what they do; and a few by what they are.[1]
   Elbert Hubbard

People have an unfortunate tendency to judge a book by its cover: a fact that can be detrimental to your career unless you know how to deal with it. This chapter shows you how to put a *cover on your book*, so to speak, that will make people want to read it. Your "cover" is your professional image—the impression you make before people get to know you well enough to understand what a capable person you are. However, before explaining how you can project a winning image, I want to preface everything presented in this chapter with the following qualifier: Image and substance are two different things. Your image, although important, should never be viewed as an alternative for substance. Image is important because it can open doors for you, but it will take substance to keep them open. You want to have both a positive image AND substance, not image instead of substance.

Think of the difference between image and substance in this way: The best looking cover in the world won't compensate for the shortcomings of a poorly written book. First, there must be substance—you must be good at what you do professionally—then, as an additional success strategy, you learn to project an image that will make a positive impression on people. There is an adage that says, ". . . you don't get a second chance to make a first impression." This is why it is so important for you to learn how to project a winning image. If you make a good first impression on people, they are more likely to give you a chance to show how smart, motivated, and capable you are. In other words, if they like the cover, they are more likely to read the book.

SUCCESS TIP

Business etiquette is not just a bunch of inconvenient and restrictive rules of behavior invented by some prim and proper grandmother with too much time on hand. Rather, business etiquette is about showing respect for the people you interact with in a business setting and for the work to be done in that setting.

In the workplace, you will often find yourself in settings where it is important to make a good impression on people you don't know. These people might be clients, customers, colleagues, or just members of the general public who will form an opinion of your organization based on their opinion of you. They will observe how you walk, talk, dress, interact with people, and otherwise comport yourself.

The fundamental skills necessary to project a positive professional image are known collectively as *business etiquette*. Business etiquette is not just a bunch of inconvenient and restrictive rules of behavior invented by some prim and proper grandmother with too much time on hand. Rather, business etiquette is about showing respect for the people you interact with in a business setting and for the work to be done in that setting. The most important rules of business etiquette fall into the following categories: greetings and introductions, dress, conversation, and meals. You need to know how to introduce yourself and others, how to dress in a way that portrays the right image, how to make conversation in both professional and social settings, how to apply table manners in any setting from a fast-food restaurant to a five-star establishment, and how to organize and conduct business meals.

GREETINGS AND INTRODUCTIONS

In order to project a positive professional image, you need to know how to introduce others, introduce yourself, shake hands properly, give your business card to potential customers, and receive business cards that are offered. This section explains how to do all of these things with just the right amount of professional élan.

Introducing Others

The key to getting introductions right is to remember that business etiquette is based on position and authority, not on age and gender as it is with social etiquette. In a business setting, always introduce the person of higher authority first. For example, the CEO of a company would be introduced first to a vice president, director, or department head. An exception to this rule is that customers and clients are always introduced first. Here are some general rules that will help you perform this task well:

■ Never ignore someone who should be introduced. If you can't remember a person's name, it's better to just say something along these lines: "I'm sorry, but I seem to be having a memory lapse. Will you remind me of your name please?" Having gotten the name, make the necessary introductions.

SUCCESS TIP

The key to getting introductions right is to remember that business etiquette is based on position and authority, not on age and gender as it is with social etiquette. In a business setting, always introduce the person of higher authority first. An exception to this rule is that customers and clients are always introduced first.

- Look at the person you are introducing as you make the introduction. Then turn and look at the other person.
- Say names slowly and clearly enough that they can be heard and understood. Also, add a conversation starter to each name. A conversation starter is just a piece of information about the person being introduced that will help the other person start a conversation or will provide a warning about topics of conversation to avoid. For example, you might say, "Tom, this is Vicky Barber. Vicky is the local compliance investigator for the Environmental Protection Agency." This tidbit of information about Vicky will help Tom begin a conversation with her, and it could give him a helpful warning of potentially touchy subjects to avoid.

When and How to Introduce Yourself

There will be many times when it will be necessary to introduce yourself in a work setting. Remember, many of the people you will interact with won't know the rules of business etiquette. Here are some situations in which you should take the initiative and introduce yourself:

- In a business or business-related social gathering when you are mixing with people you don't know and no one else introduces you to them
- When someone who should take the initiative to make introductions doesn't do it
- When you are seated next to someone you don't know in a meeting or at a business meal
- If it looks like the person who should make the introductions is having trouble remembering your name or if someone you have met in another setting does not appear to remember your name

When introducing yourself to someone, extend your right hand for a handshake, look that person in the eye and say, "Hello, my name is—. I am with ABC, Inc." This introduction gives the other person enough information to introduce himself and begin a conversation by asking, "What do you do at ABC, Inc."

Shaking Hands

Probably the most important form of nonverbal communication in a business setting is the handshake. A handshake can say so much. On the one hand, it can convey confidence, human warmth, and a friendly, open manner. On the other hand, it can convey nervousness, uncertainty, and a cold, aloof manner. It can cause people to see you as a considerate person or a domineering bully. All of these impressions and more can come from just one handshake. The difference is in how you do it. Consequently, it is critical that you learn how to shake hands in a way that conveys the right message. The following tips will help you have a handshake that conveys a positive message:

- Your handshake should be firm, but not hard. It should never cause the recipient discomfort.
- When shaking someone's hand, look that person in the eye and smile.

SUCCESS TIP

Probably the most important form of nonverbal communication in a business setting is the handshake. A handshake can say so much. On the one hand, it can convey confidence, human warmth, and a friendly, open manner. On the other hand, it can convey nervousness, uncertainty, and a cold, aloof manner.

■ Two or three pumps of the hand are plenty. Holding someone's hand too long can send a nonverbal message of implied intimacy that is inappropriate in a business setting. On the other hand, letting go too fast can convey the nonverbal message that "I don't want to touch you."

■ To avoid holding onto someone's hand too long, wait until the introduction has been made before extending your hand.

■ Shake with your right hand. If you have something in your right hand, change it to your left or put it down and then shake with your right hand. This brings up an important point. If you are in a work-related gathering, don't carry anything in your right hand. Train yourself to carry briefcases, overcoats, laptops, purses, and other items in your left hand so that your right hand is always readily available to shake.

■ Be cautious of wearing rings on your right hand that might make a handshake uncomfortable. You don't want to yelp when someone gives you a firm handshake or fall to your knees in pain when some cretin tries to prove his manhood by crushing your ring into your fingers.

■ Don't fall into the habit of using the "politician's handshake." There are three versions of this handshake—all of them bad. The first involves putting your left hand on the other person's back, bicep, or wrist while shaking with your right hand. If you are just meeting a person for the first time, patting him on the back or squeezing his bicep or wrist while shaking his hand is likely to be viewed as an obviously insincere display of inappropriate familiarity. You don't yet know the person well enough to display this level of familiarity. The second version of the politician's handshake is the two-fisted approach in which you grasp the other person's right hand in both of yours. Like the first version of the politician's handshake, the two-fisted approach conveys a level of familiarity that is out of place in

SUCCESS TIP

If you are in a work-related gathering, don't carry anything in your right hand. Train yourself to carry briefcases, overcoats, laptops, purses, and other items in your left hand so that your right hand is always readily available to shake.

a business setting when two people have just met. The two-fisted handshake should be reserved for greeting a good friend you haven't seen for a while. The final version of the politician's handshake comes into play when there are several people whose hands you should shake; for example, when you are part of a receiving line. I call this version the *grab-and-pull* handshake. With this approach you grasp the other person's hand and use it to pull him along and out of your way so you can shake the next hand. Whereas the first two versions of the politician's handshake convey insincerity and inappropriate familiarity, the grab-and-pull handshake conveys the message that "I don't have time for you—I just want to get this handshake over with."

Giving and Receiving Business Cards

Business cards are an important part of work-related discourse. They play a critical role when you want to make sure colleagues, clients, or customers know how to get in touch with you, or that you know how to get in touch with them. Your business card is a *connector*—it connects you with people you want to stay in touch with and who want to stay in touch with you. Consequently, it is important to make sure that your business card contains all of the right information. A business card should contain at least the following information:

- Your name (with your nickname or what you like to be called in quotes if it is different than your given name).
- Company name.
- Company mailing address with both PO Box and street address to accommodate both the U.S. Postal Service and the private parcel delivery companies.
- Your e-mail address and your company's Web page address.
- Your telephone number (with area code). You will have to make a decision concerning putting your cellular telephone number on your business card. There are advantages and disadvantages. Some professionals prefer to leave off their cellular number so they can control who has access to it.
- Your fax number with area code.
- Your organization's corporate logo.

When giving and receiving business cards, there are things you should do and things you shouldn't do. The most important of these are as follows:

- Discretion and subtlety are the keys when giving someone your business card. The exchange should be handled as a private transaction between two people.
- Hand your business card to the other person with the name side of the card up and turned so that the recipient can read it.
- When you receive someone's business card, silently read the person's name. Then look at the person's face and try to commit the face and the name to memory.

- It is best to wait to be asked for a business card before giving one. If you want a person to be able to get in touch with you and he does not ask for one of your cards, quietly ask, "May I give you my business card?" Don't ever just take a business card out and intrusively push it on another person.

- When you are in a group of people, avoid the temptation to give everyone a card at the same time. This will make you look more like a Las Vegas card dealer than an accomplished professional.

- Keep a good supply of your business card in an easily accessible place and in a container that protects the cards from damage. Two of the worst business-card mistakes—even worse than not having a card at all when somebody asks for one— are (1) rummaging through your wallet, purse, or pockets frantically trying to locate a card; and (2) giving someone a rumpled, stained, dirty card fished from the deepest darkest recesses of your wallet or purse. Business cards should be clean, crisp, and readily available.

APPROPRIATE DRESS IN A BUSINESS SETTING

Psychologists have long known that until we get to know people for who they really are, we judge them by how they dress, how they look (physical attractiveness), and how they talk. Like it or not, how we dress in a business setting will play a major role in determining the impression we make—at least at first. Anyone over fifty can tell you that the accepted norms of dress have changed radically in the past several decades. Business dress is much more relaxed now than it was in the buttoned-down days of the past. But one thing hasn't changed. We are still judged by how we dress. This section gives you some simple strategies that will help you dress appropriately in a business setting. The information in this section applies to both men and women.

General Rules of Dress

The most basic rules of dress have not changed in more than one hundred years. They are as follows: (1) dress for the occasion, and (2) dress for the season. For business dress, I add three more general rules: (1) dress respectfully, (2) dress nondistractingly, and (3) dress according to your organization's dress code. Dressing respectfully, means you should dress in a way that conveys the message that you respect the people you are doing business with as well as the business being done. For example, if your company is hosting a

SUCCESS TIP

The problem with *business casual* as a concept is that if the organization fails to clearly define what it means, business casual has a tendency to become *business sloppy*.

team of potential clients, and they wear business suits, it would be disrespectful of you to show up in jeans—even if jeans are the typical dress of the day in your office.

Dressing nondistractingly means dressing in a way that allows people to focus on you and the work at hand rather than on how you are dressed. Remember, being over-dressed is just as inappropriate as being under-dressed. Also, flashy and revealing outfits and gaudy jewelry are out of place in a business setting. Subtlety and conservatism work best for business dress.

Dressing according to the dress code means just what it says. The key here is to understand that there is *always* a dress code—even if it isn't written down. If your company has no written dress code, observe how the most successful managers and executives in your company dress. That's the dress code.

Specific Rules of Dress

Deciding how to dress at work is a real balancing act. On the one hand, you want to subtly stand out from the crowd in ways that promote success. On the other hand, you don't want to stand out in ways that call attention to your dress instead of your work. Here are some more specific strategies that will help you achieve the right balance when dressing for work:

- Dress like the people in your company who already have the job you would like to have. If you want to advance in an organization, identify the most successful people in the organization and emulate their dress. If you are a man and the successful person you want to emulate is a woman, find someone outside the company to serve as your role model. If you are a woman and the person you want to emulate is a man, the same rule applies.
- Dress up one notch by tending to the details. Make sure your clothes are properly ironed and accessories are in good condition. Women should also be especially attentive to the details of jewelry and hair.
- Pay attention to fit. Very few people, male or female, can buy clothes off the rack and have them fit as well as they should. Fortunately, anything bought can be tailored. Pay attention to getting the length of pant legs, skirts, dresses, and jacket sleeves just right. Also, have any excess material tapered out of shirts and blouses so they don't appear overly baggy and rumpled when tucked in.

BUSINESS CASUAL VERSUS MORE FORMAL DRESS

In many organizations, business casual is now the norm. Business casual typically amounts to wearing slacks, and an open-collared shirt embossed with the company logo. Shirts may be long- or short-sleeved depending on the season. The problem with *business casual* as a concept is that if the organization fails to clearly define what it means, business casual has a tendency to become *business sloppy.*

All too often, when a business casual dress code is adopted by an organization, people stop ironing their clothes, tucking in their shirts and blouses, and tending to normal grooming and hair care. Before long, what was intended to be just one notch down from business formal becomes just one notch up from working-in-the-yard wear. Slacks and skirts give way to jeans, polo shirts give way to tee shirts, and dress shoes give way to running shoes. Men who dress casually often overlook shaving and women often adopt casual hairstyles such as ponytails.

Clothes That Are NOT Business Casual

Organizations that adopt a business-casual dress code should define what the concept means for their employees. However, regardless of the definition arrived at by a given organization, for engineering professionals trying to build a winning career, there are certain items of clothing and approaches to dressing that should be avoided regardless of the dress code. These include the following for men and women: sweatpants, sweatshirts, athletic wear (unless you work for a company that produces it), bib overalls, shorts, tee shirts, sleeveless tops (unless worn under a jacket), any clothing item made of transparent or overly revealing material, loud colors, uncoordinated combinations such as stripes and plaids, shoes without socks (men), baseball caps, pants without a belt, blouses with plunging necklines, and short skirts and dresses. All of these various forms of casual dress have their place, but that place is not in the office.

When to Dress Up Regardless of the Dress Code

Even in this era of business casual wear, there are times when engineering professionals should dress up regardless of the dress code. These times are as follows:

- When meeting with clients who typically dress in business formal
- When meeting with potential clients or customers and you don't know how they dress
- When making a presentation on behalf of your organization to any outside group (e.g., at a professional conference, to a potential customer, to a civic group)
- When interviewing potential new employees or when being interviewed
- When conducting a performance-appraisal interview or when you are the subject of a performance-appraisal interview
- When meeting with international clients, government officials, auditors, attorneys, or board members
- When invited to represent your organization at a civic or social event and you are unsure of the dress code or when the dress code is unclear

CONVERSATION IN WORK-RELATED SOCIAL SETTINGS

The boss invites you to a social gathering in his home. You are asked to represent your organization at a chamber of commerce function. You plan to attend a national conference in your field, and one of the events will be a cocktail hour followed by a banquet.

Success Tip

> Bartenders are taught to avoid such topics as religion and politics when speaking with patrons. This is good advice for you too. In a work-related social setting, talk about good movies, good books, interesting items in the news, vacation plans, or work-related topics. But avoid topics that might arouse strong or angry reactions.

These types of activities and many others that require you to make conversation in social settings are common occurrences for working professionals. People will judge you and, through you, your organization according to how well you handle yourself in these social settings. You need to be comfortable making conversation with people from outside your profession and with people in your profession who are strangers to you. The following tips will help you become a good conversationalist in such settings.

- ■ *Avoid controversial topics that might lead to disagreements.* Bartenders are taught to avoid such topics as religion and politics when speaking with patrons. This is good advice for you too. In a work-related social setting, talk about good movies, good books, interesting items in the news, vacation plans, or work-related topics. But avoid topics that might arouse strong or angry reactions.

- ■ *Let the other person talk more while you listen.* Most people like the sound of their own voice better than the sound of yours. If you become a good listener, when you find yourself in a work-related social setting, you won't need to do much talking. Let the other person do most of the talking while you listen. Do this and other people will think you are a great conversationalist because many people would rather talk than listen. When listening to someone talk—in any setting, but especially in a work-related social setting—apply the following listening tips: (a) look the speaker in the eye and keep a friendly look on your face that conveys the message "I am listening and interested in what you have to say"; (b) concentrate on what is being said; (c) be patient—don't rush the speaker or complete sentences for him; and (d) control your emotions no matter what the other person says and no matter how much you might disagree with him.

- ■ *Use open-ended questions to get the other person talking.* An open-ended question is one that cannot be answered "yes" or "no." Open-ended questions elicit information. Consider this scenario. You want to ask the other person's opinion about new airport security regulations. You could ask, "Do you agree with the new regulations?", which is a closed-ended question because it can be answered either "yes" or "no." This would elicit a response, but it wouldn't get a conversation started. By asking the same question in an open-ended format—"What do you think of the new regulations"—you can elicit information and get the other person talking.

- ■ *Project a positive attitude when speaking.* Have a topic or two you are comfortable with always ready in case the other person needs you to do the talking. When you speak, remember to be tactful, friendly, and enthusiastic. Nobody likes to listen to

Success Tip

Certain casual eating behaviors that may be common practice when among friends having a meal in the Student Union or at a favorite college pub are not acceptable in a work-related setting.

someone who drones on in an uninterested monotone as if he is bored with the topic, the conversation, and the listener.

■ *Avoid topics that are "downers."*   You want to be remembered by those you talk to in a work-related social setting as a positive, encouraging type of person. Consequently, when it's your turn to talk, avoid topics that are downers such as the following: bad news, health problems, the high price of gasoline or anything else, the faults of other people, someone's death, etc.

■ *Don't ask questions that put people on the spot.*   The last thing you want to do when making conversation in a work-related social setting is ask the other person a question that might be embarrassing or too personal in nature. For example, never ask the following types of questions: someone's age or weight; if a woman is pregnant; where the person's spouse is (they may no longer be married); how much the person's house, car, suit, dress, watch, or anything else cost; etc.

■ *Ask the other person about positive topics.*   Not only should you avoid bringing up negative topics when it's your turn to talk, you should also avoid asking the other person about such topics. Instead, ask about positive topics such as good books recently read, movies seen, how a local sports team is doing, good restaurants, new and exciting technological developments, and anything in the news that is positive in nature.

■ *End conversations tactfully.*   When you feel it is time to end a conversation and move on to another, be sure to make the break tactfully. Don't just stop talking and walk away. The following conversation enders are both tactful and positive: (a) "It has been nice talking with you. I should go say hello to the host"; (b) "It has been a pleasure talking with you. I should see what is being served at the snack table"; or (c) "I see that John Jones of our Atlanta office is here. Let me introduce you to him" (having made the introduction, you may then exit the conversation).

Basic Table Manners

Business deals have fallen through, promotions have been denied, and salary increases have been missed all because of minor infractions relating to basic table manners. Luncheons, formal dinners, and banquets are part of doing business in every technical and business field. Consequently, it is important to learn basic table manners and how to put those manners to good use advancing your career. The table manners explained in this section apply to both work-related meals of all kinds.

Casual Behaviors to Avoid

Certain casual eating behaviors that may be common practice when among friends having a meal in the Student Union or at a favorite college pub are not acceptable in a work-related setting. Some casual behaviors to avoid and their corresponding substitute behaviors include the following:

- Blowing on food or drinks to cool them—let them sit and cool.
- Dunking food in a drink—wait until you are alone to dunk doughnuts in your coffee or cookies in your milk.
- Passing food across the table—pass food to the right.
- Crumbling crackers into your soup—eat the crackers by themselves.
- Cutting up all of your food before beginning to eat—cut just one bite at a time.
- Leaning back in your chair—keep all chair legs flat on the floor.
- Pushing your plate away when you have finished eating—leave it where it is and let the waiter pick it up.
- Picking at your teeth—excuse yourself and go to the restroom if you need to pick your teeth.
- Leaving lipstick stains on glasses—blot your lipstick before drinking.
- Talking while chewing—swallow first, then talk.
- Chewing with your mouth open—take small bites and keep your mouth closed until you have swallowed.
- Smoking during the meal or at the table. Don't. If you are a smoker, it is best to refrain completely on the occasion of a business meal. Wait until later.

It is easy to underestimate the importance of general table manners in a business setting, but don't do it. I have seen etiquette blunders in such settings cause business deals to go sour and, in turn, harm careers. I once participated in a business luncheon in which a colleague did just about everything wrong. Our CEO was at the table and this was the first time he had seen my colleague in a setting outside of the office. Bear in mind that this colleague was a talented professional who had already made good progress in advancing his career. This was why he was present during the meal in the first place—a meal with a potential customer who could bring our company a substantial amount of work. Unfortunately, during the meal my colleague proved that his career potential stopped at a level just under the point on the career ladder where you begin to meet with and entertain potential customers.

SUCCESS TIP

It is easy to underestimate the importance of general table manners in a business setting, but don't do it. I have seen etiquette blunders in such settings cause business deals to go sour and, in turn, harm careers.

My colleague made a succession of embarrassing blunders simply because he didn't know any better. In fact, later when he learned from our CEO that the potential new customer has been insulted by his lack of etiquette, my colleague was devastated. He had no idea that he had done anything wrong or out of place. But he certainly had—I was there and witnessed his blunders.

The first thing he did wrong was to pick up his napkin and snap it open with a loud "pop." Then, rather than lay the napkin across his lap, he stuffed it in his collar and let it hang down the front of his shirt. Rather than order something from the menu that would be easy, neat, and convenient to eat, my colleague ordered spaghetti and meat balls and attacked it with gusto. In the process, he splattered spaghetti sauce all over the front of his shirt—in spite of the napkin that hung there. During the meal he stuffed his mouth as full of spaghetti as possible, talked with his mouth full, and accentuated key points by jabbing the air with his fork.

After the meal, everyone at the table ordered coffee and dessert. While adding sugar to his coffee, my colleague clanged his spoon against the inside of the cup so loudly the rest of us thought he was trying to get our attention to propose a toast. Then he compounded his error by blowing vigorously on his coffee to cool it and slurping with pleasure each time he took a drink of it. My colleague topped off his list of etiquette blunders by picking several cookies out of his dessert course and dipping them in his coffee.

During his entire performance, my colleague was blissfully unaware of our CEO's growing embarrassment and our potential customer's discomfort. By the time the meal had finally ended, my colleague's image with this potential customer and our CEO had hit rock bottom. The meal wasn't a complete disaster though. The customer did eventually give our company some of his business, but with the stipulation that my colleague not be part of the team that worked on his contracts. When trying to build a winning career, you don't want your CEO to hear such stipulations about you from customers.

Basic Rules of Dining

When participating in a business meal, it is important to understand place settings; know how to properly use silverware; know how to sit and what to do with your hands; and know how to pace your eating, the proper way to butter and eat bread, and what to do if you spill something.

- *Understanding place settings.* Occasionally a work-related meal will be formal. When this is the case, understanding the place setting in front of you can be a daunting challenge. The following tips will help: (1) your place setting is bordered by your bread plate on the left and your drinking glass on the right—everything between

Success Tip

How you sit and what you do with your hands is important. Sit up straight in your chair—don't slump over your food. Do not put your elbows on the table.

the bread plate and the drinking glass is yours; (2) there will usually be three forks to the left of the dinner plate—the dinner fork has the longer tines, the salad fork the shorter tines (if there is a third, smaller fork it is a fish fork); (3) knives, spoons, and the cocktail fork (if provided) will be to the right of the dinner plate—the salad knife is usually smaller than the dinner knife and the fish knife (if provided) is yet smaller; the spoon to the right of the plate is your soup spoon; (4) there may be an additional fork and spoon above the dinner plate—if so they are your dessert silverware; (5) the salad plate will be either on the dinner plate or slightly above it and to the left; (6) your bread plate is to the left of the dinner plate and will usually have a butter spreader on it; and (7) all of your drinking glasses will be to the right and slightly above the dinner plate (water, wine, champagne flute, etc.).

■ *Proper use of silverware.* The following tips will help you use your silverware properly: (1) hold your silverware with your fingers, not your fists; (2) once you have used a piece of silverware, put it back on your bread, salad, or dinner plate as appropriate (not on the tablecloth and not partially on the plate and partially on the table); (3) don't use your silverware to emphasize key points when talking—never wave silverware around and don't point with it; (4) when cutting meat or any other type of food, pierce it with your fork and use short, gentle strokes to cut—don't appear to be sawing your food; and (5) when using a spoon, put it into the soup on the near side of the bowl and take it out on the far side—in other words, push the spoon away from you in the bowl rather than pulling it toward you.

■ *Posture, hands, and elbows.* How you sit and what you do with your hands is important. Sit up straight in your chair—don't slump over your food. Do not put your elbows on the table. The only time your hands should be above the table is when they are being used for eating or drinking. If a hand is not in use at the moment, put it in your lap.

■ *Proper pacing of the meal.* It is bad form to rush through a meal and simply sit there while everyone else is still eating. Watch others at your table, and pace your eating so that you finish at the same time as the group. One way to do this is to have the same number of courses as everyone else. If you don't plan to eat the salad, one of the entrées, or the dessert, have it placed before you anyway and just push it around on the plate a little while others are eating theirs.

■ *Proper use of bread.* Bread is a staple of most meals, and it is often one of the first items served. Eating bread at a formal meal is different from what we are typically accustomed to. Remember the following tips about eating bread during a formal meal: (1) if you plan to use butter, put some butter on your bread plate first—don't put it right on

SUCCESS TIP

It is bad form to rush through a meal and simply sit there while everyone else is still eating. Watch the others at your table, and pace your eating so that you finish at the same time as the group.

the bread directly from the butter dish; and (2) break off one small piece of the bread at a time, butter that piece, and eat it—don't butter the entire piece of bread first and then begin eating it.

■ *Handling spills and other minor crises.*   It's going to happen—it's almost inevitable. During a work-related meal, you will eventually spill something. When this happens, respond in a calm, low-key manner. Don't jump up and shout, "look out!", or make a big fuss about it. If you spill a drink, simply blot it up with a napkin and make a brief apology to others at the table while you are setting things right. If you spill gravy, wine, or sauce on your clothing, dab it off as best you can with your napkin. If you are concerned about a permanent stain, quietly excuse yourself, go to the restroom, and apply cold water to the spot.

SPECIFIC RULES FOR BUSINESS MEALS

Although both are about business, there are differences between work-related social meals and business meals. Work-related social meals are about business only in a general sense. An example of such a meal would be a chamber of commerce dinner where potential customers might be present. During such a meal, you are trying to project the best possible image for your company in the hope that new contracts might be forthcoming or that current contracts might continue to flow smoothly. Work-related social meals are more about establishing and maintaining positive relationships for yourself and your business.

Business meals, on the other hand, are meals where actual work is going to be done. At a business meal, you might discuss the details of an ongoing project or a potential new one with a customer. You might discuss progress made toward completion of a given project or problems that have arisen with the project. You might sign papers; examine graphs, charts, or drawings; look at budgets; or develop schedules. In other words, you will conduct business.

All of the general rules of dining already explained apply to business meals too. In addition to these general rules, there are some other rules that apply specifically to business meals. These rules are summarized here from the perspective of when you are responsible for setting up the meal.

■ *Remember that a business meal is a meal with a mission.*   Business meals are not social events. You do not invite a current or potential client or customer to eat with you unless you have a specific goal you want to accomplish. Arrange the meal in such a way as to help accomplish your goal.

 ## SUCCESS TIP

Remember that a business meal is a meal with a mission. Business meals are not social events. You do not invite a current or potential client or customer to eat with you unless you have a specific goal you want to accomplish. Arrange the meal in such a way as to help accomplish your goal.

■ *Select the restaurant carefully.*   Always select a restaurant you are familiar with. You don't want any unpleasant surprises to distract from the meal or the goal you are trying to accomplish with it. It is best to select a restaurant where you know the personnel, the menu, and the layout. If you are arranging a business meal in an out-of-town location, select an appropriate chain restaurant you are familiar with back in your hometown. Avoid both fast-food restaurants and five-star establishments for business meals. A good restaurant that falls in between these extremes is best for business meals. You want to appear to be neither extravagant nor cheap when hosting business meals.

■ *Take control of logistics.*   Make reservations—never leave getting the right table to fate. Arrive early—preferably fifteen minutes—and take control of the logistics. Make sure the table you have reserved is ready and appropriate (big enough, out of the flow of traffic, not near any distractions, etc.). Either give the hostess the names and descriptions of your guests or wait for them at the door. If you wait at the table, stand and shake hands with all guests when they join you. While waiting, do not order anything or even drink water. Make sure your guests sit down to a clean, unused table.

■ *Organize the seating yourself.*   Either make seating cards beforehand that show who sits where, or do the seating yourself. It is important that you be seated in a position across from the key decision maker in the client's group. You want to be able to look this person directly in the eyes, and you want to be able to talk with others present without turning the back of your head to the key decision maker.

■ *"Conduct" the meal.*   You organized the meal. Consequently, you are in charge. When in charge, take charge. Your guests will watch you for their cues. The host is supposed to be the first person to place the napkin in his lap—so do it. The others will follow your lead. Simply pick it up, unobtrusively unfold it, and put it in your lap. Then, enjoy your meal. When you sense that all of your guests have finished eating, place your napkin loosely on the table. Don't try to refold it. This is the sign that the meal is over and business can begin.

■ *Order alcohol with caution, if at all.*   The server will typically ask for drink orders before taking orders for meals. When the server asks if you would like a drink, defer to your guest. If the guest orders wine or any other alcoholic drink, you may order one too. However, if your guest does not order an alcoholic drink, you shouldn't either. You do not have to order an alcoholic drink even if your guest does. However, be tactful in how you say "no" in this case. Never say "I don't drink" or anything else that might make your guest uneasy. Rather, say "I don't think I'll have a drink right now, but please feel free to have one yourself." If wine is ordered, the server will present you with the cork before pouring. Simply pinch the cork to make sure it is damp. Don't smell it and, for sure, don't lick it. If the cork is damp, nod to the server and he will pour a small amount of wine in your glass. Taste the wine. If it is acceptable, nod to the server. He will first fill your guest's glass and then yours. If the wine is not acceptable simply ask for another bottle or another kind.

■ *Order your food with confidence, but discernment.*   One of the reasons you choose a restaurant well known to you is that when it's your turn to order, you know the menu well enough to order quickly and with confidence. Taking a long time to order, struggling to make up your mind, or making a grand production out of ordering will give your guest

SUCCESS TIP

Most of what you have seen people do in movies and on television when making toasts is incorrect. For example, if you wish to propose a toast to your guest, you don't clink a knife or fork against a glass to get everyone's attention. Rather, simply stand up and clear your throat. Then wait for the conversation at the table to stop.

the wrong impression. In addition, make sure that what you order is easy to eat and not messy. You don't want your guest watching in embarrassed amusement as you wrestle with a three-inch thick sandwich that falls apart every time you try to take a bite, or as you try in vain to dodge the dripping tomato sauce from your spaghetti.

■ *Propose toasts with care.*   There will occasionally be a reason to propose a toast during a business meal. Most of what you have seen people do in movies and on television when making toasts is incorrect. For example, if you wish to propose a toast to your guest, you don't clink a knife or fork against a glass to get everyone's attention. Rather, simply stand up and clear your throat. Then wait for the conversation at the table to stop. Keep your toast brief, to the point, and complimentary of the guest you are toasting. Never write out toasts and read them. If a toast is too long to memorize, shorten it until you can memorize it. Even if you have an excellent memory, keep toasts short. Also, it is bad form to clink glasses after a toast. The proper action is to raise your glass, not to clink it with someone else's. Clinking is a good way to break a glass and cause an embarrassing spill—or worse. I was once present at a business meal where everyone clinked glasses after a toast. The glass of the guest being toasted broke, spilling wine down the front of his suit. As a result both the meal and the deal went sour. By the way, you may toast someone using any liquid—water, tea, and coffee are just as appropriate as wine. If you are the host, you should toast first. If you are the guest, wait for the host to exercise his privilege before proposing a toast.

■ *Order dessert with tact.*   When the server asks if you will be having dessert, defer to your guest, and, at the same time, encourage your guest to order dessert. This takes the onus off the guest. If your guest orders dessert, you should too—even if you don't want it. In such a case, just order it and push it around on the plate with your fork as your guest eats his. You don't want your guest to feel guilty eating dessert while you abstain.

■ *Remember your manners with coffee and tea.*   Coffee or tea can provide a good ending to a meal. In fact, a good time to relax a little and talk business is while drinking coffee or tea. But don't overdo it. Relax, but don't get sloppy. Coffee and tea are just as much a part of a business meal as the main course. When stirring your coffee or tea, be careful that you don't clank the spoon against the sides of the cup or swirl so vigorously that the contents spill over the sides onto the saucer. Once you have stirred in the sugar and cream, put the spoon on the saucer—don't lick it. Sip coffee and tea silently—don't slurp. Never blow on coffee or tea no matter how hot. If your drink is too hot, leave it alone for a few minutes. This is important. If you take a big

SUCCESS TIP

> When stirring your coffee or tea, be careful that you don't clank the spoon against the sides of the cup or swirl so vigorously that the contents spill over the sides onto the saucer. Once you have stirred in the sugar and cream, put the spoon on the saucer—don't lick it. Sip coffee and tea silently—don't slurp. Never blow on coffee or tea no matter how hot. If your drink is too hot, leave it alone for a few minutes.

gulp of coffee or tea that is too hot, you are going to have several moments of intense discomfort followed by watery-eyed embarrassment.

- *Pay the bill if you are the host.* There should never be any haggling over who pays the bill for a business meal. If you are the host, you pay. Discretion is the rule when paying the bill. Bills are best paid in private. One way to do this is to wait until you have escorted your guest to the door before paying. Another even better approach is to arrive early for the meal and give the cashier a credit card imprint and arrange things so that the bill never comes to the table. Then, if your guest attempts to pay all or a portion of the bill, you may simply say, "It's already been taken care of."

Remember, business etiquette is not about putting on airs. It's about making the right impression—an impression that is good for you and your business. It's also about showing an appropriate level of respect for your customers and the business you do with them. Finally, it's about advancing your career.

REVIEW QUESTIONS

1. In a business-related social setting, when you are not introduced, it might be necessary to introduce yourself. Describe several situations in which introducing yourself is appropriate.
2. If you want to be sure that you make a positive impression when shaking hands, what are some things you should remember? Name at least five.
3. Describe the information that should be on your business card.
4. Explain at least four rules of thumb for properly giving someone your business card.
5. Explain three broad rules of dress you should follow in a business setting.
6. What is the fundamental weakness of the business casual dress code?
7. List several items of clothing for your gender that are NOT business casual.
8. Explain several situations in which you should dress more formally regardless of the prevailing dress code.
9. Explain several rules of thumb you can use to advantage in conversations that take place in business-related social settings.
10. List at least three behaviors to avoid at business meals.
11. Explain how to properly pace yourself during a business meal.
12. Summarize the basic specific rules for business meals.

DISCUSSION QUESTIONS

1. Sandy Brown isn't sure what to do. She has just joined a group of potential customers who are talking to her boss, Penny Morris, as they have cocktails and wait for others to arrive for a business meal. Without introducing Brown, Penny Morris walks away and joins another conversation. Discuss how Sandy Brown should handle this situation.

2. Peggy Andrews is in a quandary. She is seated in a restaurant waiting for a client to join her for a business lunch. The manager of the restaurant finally tells her that the client called earlier with his apologies. He will not be able to join her for lunch, but asks that she leave her business card with the restaurant manager. The client will pick the card up later and get in touch with her to reschedule. Andrews is uncomfortable leaving a business card that contains her address, telephone number, fax number, e-mail address, and cell phone number with a stranger. Discuss how Peggy Andrews might handle this situation.

3. You have been invited to attend a meeting with a new client you have never met. You have no idea of how the client dresses, but you want to make a good impression. How should you dress? Discuss the possibilities.

4. Discuss the following statement. Business people used to dress to make a positive impression on customers, colleagues, and associates. Now they just dress for personal comfort and don't care what kind of impression they make. Should the impression you make on others be a factor in choosing a dress code?

5. Discuss the following situation. You are seated at a business meal with a potential customer who is tight-lipped and doesn't say much. How might you get this person to open up and talk?

6. At a business meal, you spill your drink on the table. Discuss how you should handle the embarrassing situation.

7. Your boss says, "We have an important client coming in tonight. I'm going to take him to the best restaurant in town. I hope it's as good as they say it is. I've never been there." Do you see a problem here? Discuss what your boss should do.

ENDNOTE

1. Louis E. Boone, *Quotable Business*, 2nd ed. (New York: Random House, Inc., 1999), 115.

CHAPTER NINE

Become an Effective Negotiator

Let us never negotiate out of fear. But let us never fear to negotiate.[1]
   John F. Kennedy

So much of what happens in the workplace involves negotiating. You might be part of a team that negotiates a new contract for your company, or you might have to negotiate with your boss for a raise. You might need to negotiate with team members to determine who has to work late, or you might have to negotiate the price your organization will pay for a large order of supplies. You might need to negotiate various job-related issues such as childcare, flex time, contracts, salary, bonuses, and expense accounts. Even deciding among friends where to go for lunch can be a negotiation.

   The most successful people are typically good negotiators. They have to be in order to succeed. This is not to say they are born as good negotiators. Negotiating is a skill that can and should be learned. Like any skill, the more you practice, the better you become at using it. The word *become* is important here because being a good negotiator is not something you *are*, it's something you *become*. What you need to know in order to become a good negotiator is presented in this chapter.

NEGOTIATION DEFINED

I begin with a definition of the concept because it is important for you to understand what *negotiation* means before beginning to develop your negotiating skills. This is important because television and movies often portray the concept as meaning to outmaneuver the other side in ways that give you all the value while leaving your hapless opponent empty-handed and wondering what happened. Let me say at the outset that this view of negotiating is both inaccurate and shortsighted.

   I define the concept of *negotiation* as follows: *two partners working together to transact business in ways that are mutually beneficial and that leave the door open for future business transactions between the partners.* Although this appears to be a simple definition, it is loaded with meaning. The term *partner* is used to make it clear that the two sides in a business negotiation should not be enemies, adversaries, or even opponents. The reason for this is found in the "mutually beneficial" element of the definition.

Success Tip

A good negotiation will result in a deal that is fair and balanced so that both parties to the deal benefit. In other words, the final deal is mutually beneficial. People who go into negotiations seeking a result that is lopsided in their favor are violating the key elements of the definition. Any deal negotiated today should lay the groundwork for future deals with the same partner tomorrow.

A good negotiation will result in a deal that is fair and balanced so that both parties to the deal benefit. In other words, the final deal is mutually beneficial. People who go into negotiations seeking a result that is lopsided in their favor are violating the key elements of the definition. Any deal negotiated today should lay the groundwork for future deals with the same partner tomorrow. Repeat business is critical in a competitive environment, and especially important to engineering firms. Negotiating a lopsided deal is the fastest and most effective way to undermine future business with the *victim* of the deal. Consequently, negotiating for a lopsided result can mean that you "win" in the short term, but lose in the long term. When a negotiation is concluded, there should be no victims.

Characteristics of Good Negotiators

If you've ever watched a movie in which a negotiation takes place, chances are you saw the protagonist portrayed as a slick dealmaker who outfoxes the outclassed victim and pulls the rug out from under him at the last moment with some brilliant but devious strategy. While this might make for a good movie, what it portrays is bad business. Good negotiators are not slick-dealing, domineering, cigar-chomping, macho types who overpower their witless opponents; nor are they indecisive, submissive, unimaginative types who are afraid of their own shadows. Rather, good negotiators are typically patient, fair, well informed, cooperative, innovative, imaginative, and intuitive. These are the characteristics that are most likely to lead to deals that are fair, balanced, and mutually beneficial.

People who develop these characteristics and apply them during negotiations will typically also display the following behaviors:

- Quickly see through the fog of debate to the heart of a matter.
- Solve problems in real time before they can derail the process.
- Think clearly under pressure.
- Depersonalize comments that are made and see through emotionally charged language to the issues in question.
- Listen carefully and with patience.
- Approach the process with an open mind knowing there is always more than one way to achieve a desired result.

- Develop alternative options and solutions quickly and on the spot.
- Watch for verbal and nonverbal cues and use them to assess people.
- Think critically (i.e., recognize assumptions, rationalizations, justifications, and biased information that are presented as facts).
- Take the long-term/repeat business perspective.
- Give themselves and their negotiating partner room to maneuver (i.e., don't box people in).
- Maintain a sense of humor and a positive attitude throughout the process.
- Consider issues from the other side's point of view as well as your own.
- Understand that timing is important to success in negotiating.
- Prepare, prepare, prepare.

PREPARATION IS KEY WHEN NEGOTIATING

The behaviors explained in the previous section are all important to successful negotiations. But none is more important than the last one—preparation. Going into a negotiation unprepared will almost guarantee a bad result. The following questions will help ensure that you are well prepared for a negotiation:

- What do we want out of this negotiation? What does the other side want?
- What are we willing to give up in the negotiation in order to get what we want? What is the other side willing to give up?
- What is at risk here for us and for them? In other words, if the deal falls through, what do we lose and what do they lose?
- How much do we know about the other side and their needs? How much do they know about us?
- Is there anyone on their team who might be an advocate for us? Is there anyone on our team who might be an advocate for them?
- Do we have any "hot-button" issues? Do they?
- What don't we know about the other side, and who can help us learn what we don't know?

SUCCESS TIP

Good preparation is essential to a successful negotiation. Before beginning a negotiation you must know what you want and what you don't want, what you can accept and what you cannot accept, what you are willing to give up in order to get what you want and what you are not willing to give up. It is also important for you to understand that there is more than one way for you to get what you want.

■ Are there factors that might affect the outcome that we or they have no control over? What are those factors?

■ What is our bottom line—at what point do we just walk away? What is their bottom line?

■ What are the easy issues we can use to generate early agreement? What are their easy issues?

Good preparation is essential to a successful negotiation. Before beginning a negotiation you must know what you want and what you don't want, what you can accept and what you cannot accept, what you are willing to give up in order to get what you want and what you are not willing to give up. It is also important for you to understand that there is more than one way for you to get what you want.

The story of Terry illustrates this last point. Terry was a mechanical engineer and, as the result of a divorce, was also a single mother. Childcare costs were making it difficult for Terry to make ends meet. Consequently, she arranged a meeting with her supervisor to ask for a raise. The supervisor respected Terry, admired her work, and was sympathetic to her plight. Terry was an excellent engineer, and the supervisor wanted to help her. Unfortunately, his hands were tied when it came to giving Terry a raise.

The company they worked for had a salary schedule that paid engineers based on their level of education, years of experience, and performance. Terry's performance was excellent. Consequently, she was scheduled to get her next raise in about a year. At the time in question, Terry had just been promoted to a new level; meaning she had recently gotten a raise. The company's policy was that an employee had to perform well at a given level for at least one full year before receiving the next raise.

The supervisor told Terry that if she could make it for another eleven months, she was virtually guaranteed the raise she wanted. However, Terry was adamant she needed another raise immediately. In an attempt to solve the problem without violating company policy on raises and promotions, the supervisor offered Terry a counterproposal. The company had a new program that would help pay the childcare costs of employees who qualified for it. If Terry would accept this program instead of a raise, her childcare costs would be reduced by an amount almost equal to the raise she was demanding. But Terry was so intensely focused on the raise that she failed to see compensation for childcare costs as an equally viable solution. In anger and frustration, Terry quit.

SUCCESS TIP

When they first get involved in a negotiation, most people want to jump right in and make the deal. Closing the deal is actually the third stage in a process that has three distinct stages. A good rule of thumb is to go through each stage, even if a given stage takes only a minute or two. This is because the first stage sets up the second and the second sets up the third. If you skip a stage, you are likely to find yourself going backwards in the negotiation just when you have made a case for moving forward.

Her resignation was a major blow to the company's engineering department, but it turned out to be an even bigger blow to Terry. Having given up her only source of income, Terry was eventually forced to accept a position with another company that paid even less than the one she had before quitting. There were better job offers, but all of them required relocating to another state—something she could not do. Terry needed a local job because her divorce decree didn't allow Terry to relocate her children farther than a hundred miles from her former husband's home.

This is an example of why it is so important to understand what you really need out of a negotiation. Terry thought she needed a raise when, in fact, what she really needed was financial relief. As it turned out, accepting the offer of a company-provided childcare subsidy would have benefited Terry even more than the raise she demanded after tax considerations were factored in.

CONDUCTING NEGOTIATIONS

Once you have prepared yourself to negotiate, there are strategies you can apply to help the process work better. These strategies have to do with understanding the stages of the negotiating process, timing, location, image, creating favorable momentum, and behavior during the actual negotiation. Strategies in each of these categories of concern are presented in this section.

Negotiate in Stages

When they first get involved in a negotiation, most people want to jump right in and make the deal. Closing the deal is actually the third stage in a process that has three distinct stages. A good rule of thumb is to go through each stage, even if a given stage takes only a minute or two. This is because the first stage sets up the second and the second sets up the third. If you skip a stage, you are likely to find yourself going backwards in the negotiation just when you have made a case for moving forward.

I use the *bridge analogy* to describe the stages in a negotiation. At the beginning of the process a river runs between you and your negotiating partner that keeps you apart. You are on one bank and your negotiating partner is on the other. In order to get together, you must build a bridge. Stage one in the process involves building the foundation (pilings, girders, and beams). Stage two involves building the skeletal structure for the bridge on top of the foundation. Stage three involves adding the finishing touches. Once stage three has been completed, you and your negotiating partner can come together at the middle of the bridge.

Based on this analogy, I label the three stages in the negotiating process as follows: Stage 1—Building the foundation; Stage 2—Building the structure; and Stage 3—Completing the bridge. What occurs in each of these stages is explained in the following paragraphs.

Stage 1—Building the Foundation

Building the foundation involves laying the groundwork for the negotiating process. This is the stage in which you convince your negotiating partner that he needs to hear what

Success Tip

Timing and location are important factors in the negotiating process. For example, you don't want to find yourself negotiating the price of a new house the day before you have to be out of your old house. Make sure you select a time that is advantageous so that neither side is rushed.

you have to say. In other words, you state your case explaining how a successful negotiation will have mutual benefits. If you cannot successfully complete this stage, there is no reason to go to the next. Both parties must see the need for negotiating before proceeding with the remainder of the process. For this reason, it is important to practice explaining the need for negotiation in terms of how the other party will benefit.

Stage 2—Building the Structure

Once it is apparent that your partner understands the need to negotiate, you can get into the specifics. These include your expectations as well as those of your negotiating partner, an explanation of how your partner will benefit from what you are proposing, and specifics concerning your needs and those of your partner. If you were buying a car, this would be the point in the process when you set forth the details of what you want (e.g., color, features, size). This is not the step in which price is discussed. When both parties have agreed on the details, then you move to the final stage.

Stage 3—Completing the Bridge

At this point both parties have agreed that they would like to make a deal and both their needs and expectations have been explained. The final step in this stage—the one most often associated with negotiating—involves bargaining over price, delivery dates, warranties, and other negotiable factors. This step will be much easier to conduct if you have done a thorough job in stages one and two.

Select the Time and Place with Care

Timing and location are important factors in the negotiating process. For example, you don't want to find yourself negotiating the price of a new house the day before you have to be out of your old house. Make sure you select a time that is advantageous so that neither side is rushed. The best negotiations allow both sides plenty of time to consider offers, options, proposals, and counterproposals.

The best location for a negotiation is a neutral site. Having the negotiation take place in the other party's facility can put you at a disadvantage and, of course, the opposite is also true. Giving either negotiating partner the "home-field advantage" can be a mistake. From your perspective it might amount to giving your partner an advantage over you during the negotiating process. From your partner's perspective it might create resentment and rule out any future business between the two of you.

Project an Advantageous Image

The impression you make on your negotiating partner is critical. People are typically more willing to deal with someone they can relate to. Consequently, it is important for you to do the research necessary to know your negotiating partner before the negotiating process begins. What are the core values of this person or organization? How do people in your partner's company dress? What buzzwords are part of your partner's corporate culture?

Much of this information can be determined by locating and studying your partner's Web site. Core values of the organization can be found in its strategic plan, which is typically posted on the organization's Web site. If you know the person or people who will comprise your partner's negotiating team, look for biographies on their company's Web site or conduct an online search keying in their names. Your partner's language tendencies will show up in its vision and mission statements and in other documents contained in its Web site. You can also identify others who have dealt with your negotiating partner and question them concerning values, dress, and language.

With what you learn from researching your negotiating partner, you can decide how to dress during the negotiation, value-laden statements to espouse and to avoid, and what buzzwords to drop during conversation. There is a sense in which negotiating is like acting out a role in a play. You want to make the right impression on the audience; in this case the audience is your negotiating partner.

I once helped prepare the negotiating team of an engineering firm that was trying to land a contract with a retail chain to design and build a series of stores. Against my advice, this engineering firm's negotiating team came to the first meeting dressed casually. I had advised the engineering firm to determine how the retailer's representatives typically dressed and to be guided by what was learned. The representatives of the retail chain came to the meeting in business formal wear. In addition to neglecting the clothing issue, the representatives of the engineering firm made their proposals in the terminology of engineers rather than adapting their language for the understanding of lay people. The engineers simply ignored the dress, values, and language of their potential client—the retail chain. As a consequence, the negotiations got off to a bad start and quickly went downhill from there. The contract eventually went to another engineering firm.

Create Favorable Momentum

Baseball teams try to score at least one run in their first time at bat. Football teams try to score a touchdown on their first possession. Tennis players try to win the first game or break the other player's serve during the first game. In all of these examples, the goal is to

SUCCESS TIP

Momentum is the impetus or tendency of something to go in a certain direction. If you can get the negotiation going in the right direction from the outset, it will tend to keep going in the right direction the rest of the way. You want momentum working in your favor from the outset.

create favorable momentum. Momentum is the impetus or tendency of something to go in a certain direction. If you can get the negotiation going in the right direction from the outset, it will tend to keep going in the right direction the rest of the way. You want momentum working in your favor from the outset.

Additional Strategies for Use During the Negotiation

Once the negotiation has commenced and seems to be moving along, you want to keep it going in the right direction. Momentum gained can be quickly lost if you fail to do what is necessary to keep the ball rolling. The following strategies can be used during the negotiation to keep things moving in the right direction:

- *Think critically.*   Don't confuse facts with opinions or issues with positions. Tactfully insist on facts to back opinions and be quick to point out that issues are not positions. Issues can be resolved.

- *Listen to what is said and what is not said.*   Unfortunately, your negotiating partner might not begin the negotiation as a partner. You might have to bring him along and turn him into a partner. Consequently, you should listen attentively not just to what is said, but also to what is not said. If you sense that the other party is holding back, withholding information, or putting a certain "spin" on his proposals and counterproposals, this could be evidence of a hidden agenda. Don't be afraid to tactfully say, "Something seems to be missing in the discussion. Can you add any information beyond what you've said to clarify things?" It might take you a while to earn sufficient trust to convince the other party to drop his guard and be a partner. But if he is negotiating in good faith, he will eventually see that you are being honest and will follow your lead. If he is not negotiating in good faith, you don't want to deal with him anyway.

- *Keep your partner's needs and hopes in mind.*   When preparing for the negotiation, you came to some conclusion about what your partner wanted to achieve by negotiating. Keep your partner's needs and hopes in mind during the process. Before you make a proposal or a counterproposal, ask yourself how it might affect your partner's needs and hopes. Can you make your proposal in a way that will achieve your goal or a sufficient enough part of it while still meeting an equally sufficient portion of your partner's needs?

- *Be patient.*   Don't try to rush negotiations. This will only serve to raise suspicion in the mind of the other party. Be patient. Give things time to develop. Remember, you can apply this principle only if you have scheduled the negotiation early enough that you don't need to rush. Consequently, keep scheduling in mind when arranging the timing of negotiations.

SUCCESS TIP

Avoid stating bottom-line positions. Stating your bottom line can back you into a corner, and make you look foolish if, after stating a bottom-line position, you are forced to change positions.

SUCCESS TIP

Whatever you promised to do during the negotiation, make sure that you now follow through and do it, both properly and on time. If unanticipated problems arise, make your partner aware of them immediately. The trust and credibility you developed during the negotiation will now be either reinforced or lost based on how well you perform.

■ *Ignore personal comments.* Occasionally your negotiating partner might make a comment you find offensive. Just bite your tongue and ignore the personal aspects of the comments. Be objective and don't take things personally. It is likely in these cases that the other party is just a poor negotiator and doesn't know how to make proposals or counterproposals without getting personal. However, it could be that the other party is trying to agitate you as a way to test you or to gain an advantage. Stay calm, depersonalize, and stay focused. Negotiators who use personal remarks to gain advantage are trying to break your focus. Once they see that their tactic isn't working, they will drop it.

■ *Leave yourself room to maneuver.* Avoid stating bottom-line positions. Stating your bottom line can back you into a corner, and make you look foolish if, after stating a bottom-line position, you are forced to change positions. There are many different ways to solve the same problem or meet the same need. Although you have attempted to anticipate these various approaches, you might have overlooked one or two. Consequently, it is wise to remain open to a better idea the other party might propose.

AFTER MAKING THE DEAL—FOLLOW THROUGH

The negotiation process does not end once the deal has been made. As soon as you conclude one negotiation, you should begin paving the way for a future negotiation with the same party. Remember, negotiations have more than one purpose. The first, of course, is to make the deal in question at the moment. The second is to generate additional future business with your new partner. The best way to achieve the second purpose, assuming you have concluded a deal that is good for both parties, is to follow through and perform effectively.

Whatever you promised to do during the negotiation, make sure that you now follow through and do it, both properly and on time. If unanticipated problems arise, make your partner aware of them immediately. The trust and credibility you developed during the negotiation will now be either reinforced or lost based on how well you perform. Stay in touch with your partners, keep them informed, solicit feedback, and always be available and responsive to them. The better you perform on the current contract, the easier the negotiations will be for future contracts.

REVIEW QUESTIONS

1. Use your own words to define the term *negotiation.*
2. List ten characteristics of a good negotiator.
3. Explain the importance of preparation in negotiating.
4. List and explain the various stages in a negotiation.
5. Explain the importance of selecting the right time and place for negotiations.
6. How can you make a positive impression on the other side during negotiations?
7. Explain how to create favorable momentum when negotiating.
8. List and explain at least five strategies you can use to keep negotiations moving in the right direction.
9. Explain the concept of *follow through* as it relates to negotiations and why it is so important.

DISCUSSION QUESTIONS

1. Assume you have been asked to chair a negotiating team for your company. Now you must choose your team members. One of your colleagues has a wealth of knowledge in the areas to be negotiated, but also has several characteristics that could cause problems. For example, he has a bad temper and tends to lose his temper when people disagree with him. Should you include this person on the negotiating team? Why or why not? If you do not include him, is there any way to take advantage of his expertise?
2. Assume your company has won a new contract, but the details still have to be negotiated. You are asked to chair the negotiating team. Discuss the various things you would do to prepare for the negotiations.
3. Whenever Amy Stanford is involved in a negotiation, she goes straight for the bottom line with no preliminaries. Discuss Stanford's approach. How successful do you think she will be as a negotiator?
4. Assume you are chairing your company's negotiating team and things are going well. Both sides have taken a break for lunch. When the negotiations resume, you want to keep the positive momentum going. Discuss some strategies you can use to do this.

ENDNOTE

1. Louis E. Boone, *Quotable Business,* 2nd ed. (New York: Random House, Inc., 1999), 113.

CHAPTER TEN

Apply Self-Discipline and Time Management

All great leaders have understood that their number one responsibility was for their own discipline and growth. If they could not lead themselves, they could not lead others.[1]

John C. Maxwell

Andy has a good job with a growing engineering firm, and he is good at what he does. As the company grows, Andy hopes to grow with it. But he has a problem. Andy is a poor manager of his time. Consequently, he often wastes both his time and the time of others. Andy knows his poor time management might limit his potential for advancement, and he wants to do better, but he just does not seem to be able to break old habits. Andy's problem is that he lacks an important skill—self-discipline.

Many people decide to diet, but few stick to it. Many people start on an exercise program, but few work out regularly. Many people make New Year's resolutions, but few keep them. In each of these examples, the missing ingredient is self-discipline. Once started on a course of action, whether it is a diet, exercise program, career, or any other endeavor, you will need self-discipline in order to stick to it.

Those who fail to follow through on plans, programs, and commitments, often rationalize their failure by saying, "*I just don't have any self-discipline.*" In reality, people who use this excuse do in fact have self-discipline, they just don't apply it. Self-discipline is not something one either has or doesn't have. It is neither a gift nor a genetic trait. To say "I have no self-discipline" is like saying "I can't play tennis." Like tennis, self-discipline is a skill—something that can be learned and, once learned, must be practiced

SUCCESS TIP

Self-disciplined people are just as tempted to take the easy, comfortable, or expedient way as any one else. The difference is that they consciously choose to make the right decision and do the right thing. Self-disciplined people are self-disciplined because they choose to be.

Success Tip

Often the key difference between success and failure is self-discipline. The pathway to failure is paved with the unmet potential of talented people who failed because they lacked self-discipline. This is why self-discipline is so important to those who want to build a winning career.

consistently. Engineering professionals who hope to have winning careers must develop and apply the skill of self-discipline. This point cannot be over-emphasized.

Self-Discipline Defined

Self-discipline is the ability to consciously take control of your personal choices, decisions, actions, and behavior. Two key aspects of this definition are found in the terms *ability* and *consciously*. The term *ability* is important because it conveys the message that self-discipline—like any other ability—can be learned and then developed through consistent practice. The term *consciously* is important because it conveys the message that self-discipline is a *choice*, rather than something one just has or doesn't have.

People who exercise self-discipline are just as susceptible to the shortcomings of human nature as anyone else. The difference between those who do and those who don't exercise self-discipline is *choice*. Those who do not choose to exercise self-discipline tend to think that making the right choice in a given situation is somehow easier for other people. But this is not true. Self-disciplined people are just as tempted to take the easy, comfortable, or expedient way as anyone else. The difference is that they consciously choose to make the right decision and do the right thing. Self-disciplined people are self-disciplined because they choose to be.

Exercising self-discipline might amount to nothing more than getting out of bed when you feel like sleeping a little longer. It might involve controlling your temper when you would really like to tell someone what you really think of them. It might mean staying behind to finish the work on an important project when good friends have invited you to a party. It almost always means choosing Option B when Option A would be easier, more comfortable, more fun, or more expedient, but wrong.

There is always an element of self-denial present in the application of self-discipline. Often the principal barrier to doing what is right when another option is more appealing is that selecting the right course of action will mean self-denial. People who will not deny themselves the luxury of the easy, comfortable, or expedient option have not yet developed self-discipline.

Why Self-Discipline Is Important

Often the key difference between success and failure is self-discipline. The pathway to failure is paved with the unmet potential of talented people who failed because they lacked self-discipline. This is why self-discipline is so important to those who want to build a winning career. Self-discipline manifests itself in several different ways in a work setting, all of them positive. Consider the following benefits of self-discipline.

Time Management

In order to succeed at work, you will have to make efficient use of time. The modern workplace is a hectic, busy environment. Consequently, time management is important, and good time management requires self-discipline. Self-disciplined professionals make effective use of their time, which in turn improves their performance. Improved performance, in turn, leads to career advancement.

Stewardship

The most successful engineering professionals are good stewards of the financial, technological, and human resources entrusted to them. Overseeing all of these various resources while still tending to the everyday aspects of the job (paperwork, meetings, e-mail, telephone calls, etc.) takes time. Because self-disciplined professionals manage their time well, they have more time to devote to being good stewards. This is important because good stewardship will help advance your career.

Execution

Self-disciplined engineering professionals are typically more efficient and effective in executing their plans. Even the best plans in the world are just dreams until they are executed. Effective execution requires self-discipline. This is critical because in a competitive environment the difference between winning and losing is often how effectively plans are executed. Engineering professionals who earn a reputation for being able to execute plans go farther faster in their careers than those who can plan but cannot execute.

Followership

Self-disciplined engineering professionals are better able to do the things necessary to win and maintain followership (e.g., consistently set a positive example, maintain a can-do attitude, establish and maintain credibility). This is because self-disciplined professionals will do these things even on days when they don't feel like it. Self-discipline is fundamental to becoming a leader because it helps establish and maintain followership and those who become good leaders go farther in their careers.

SELF-DISCIPLINE: AN EXAMPLE

The value of self-discipline can be seen in the following scenario involving two engineering professionals. Mary and Mark were department heads for the same company. Although they were the same age, graduated from the same university with the same degree, and had the same amount of work experience, their departments performed at radically different levels. Mary's department typically had the highest performance rating in the company. Mark's department, on the other hand, typically had the lowest. In spite of the many similarities between Mary and Mark, these two professionals differed significantly when it came to self-discipline.

Mary was an excellent time manager. Whenever she conducted a meeting, there was always an agenda with both a starting and a projected ending time. In addition, all participants received the agenda—along with backup material—at least twenty-four hours before the meeting convened. This approach minimized the amount of time wasted in Mary's meetings. Mark, on the other hand, was known for conducting meetings that were long on time, but short on organization. He did not use an agenda, and frequently left participants idly waiting while he exited the room to search for backup material he forgot to bring to the meeting.

Mary was a good steward of the resources for which she was responsible. She saw to it that her direct reports received the training and mentoring they needed to do their jobs and to grow professionally. Because she was a wise and careful budget manager, Mary was able to keep the technologies in her department up to date and operating at maximum efficiency. Mark, on the other hand, released his direct reports for training only when it was absolutely necessary. When it came to budget management, Mark's philosophy was "... I'm not an accountant." As a result, Mark's departmental budget often ran dry before the end of the fiscal year. He did try to keep his department's technologies up to date, but because he was a poor budget manager, he seldom had the necessary funds.

Mary's time management and good stewardship were direct results of her self-discipline. Mark's poor time management and weak budget management were the results of a lack of self-discipline. Mark could have taken the time to develop an agenda and organize backup materials for meetings as Mary always did, but he chose not to. Mark could have monitored his departmental budget on a daily basis as Mary did, but he chose not to. As a result of her self-discipline, Mary had a much more successful career than Mark.

ASSESSING YOUR SELF-DISCIPLINE

Are you a self-disciplined person? A good way to answer this question is by conducting a self-assessment. Here are some questions you can ask yourself to assess your self-discipline. The goal is to be a perfect ten.

1. Do you consistently arrive at class on time or early?
2. Do you consistently arrive on time for meetings, appointments, and other similar obligations?
3. Do you consistently submit assignments on time or early?
4. Do you consistently keep up with other school-related responsibilities?
5. Do you consistently keep up with reading in your college courses?
6. Do you consistently and promptly return telephone calls?
7. Do you consistently and promptly return e-mail?
8. Do you consistently begin student meetings you chair on time?
9. Do you consistently end student meetings you chair on time?
10. Do you consistently and promptly follow through on promises?

These questions represent minimum-level expectations. Self-disciplined people would be able to answer "Yes" to all ten questions, giving them a *self-discipline quotient* of ten. Any question answered "No" indicates a need to improve your self-discipline. The functional word in each question is *consistently.*

A "Yes" answer to a question means that you do what the question asks day after day, week after week, and month after month. This does not mean that you are never late or that you never let a meeting run over time. Even the most self-disciplined person could make no such claim because nobody has sufficient control over events and circumstances to enjoy a perfect record. After all, something as unpredictable and uncontrollable as a traffic jam can make you late. And in meetings, running over time might be warranted if the topic of discussion is important enough. What a "Yes" answer to any of the questions does say, however, is that "I do this except in the most rare of instances." With this clarification, a self-disciplined person should be able to answer "Yes" to all ten of the questions.

REVIEW QUESTIONS

1. In your own words, define the concept of *self-discipline.*
2. Explain why self-discipline is so important for those who want to build a winning career.
3. Explain the concept of *choice* as it relates to self-discipline.
4. List several behaviors that could be evidence of self-discipline or the lack of it.
5. How does self-discipline relate to the concept of *execution*? Why is this concept important to those trying to build a winning career?

SUCCESS PROFILE

Stanley B. Davis
Engineering Executive and Quality Consultant

The career of Stanley B. Davis exemplifies the concept of *self-discipline* as well as several of the other success strategies recommended in this book. Davis was already working in the field of engineering before he even began his college studies. As a telemetry engineer for RCA Corporation, Davis was assigned to a project that supported NASA's Gemini and Apollo space flight programs when he realized that future promotions might require a college degree.

With a family to support, there was no way that Davis could simply quit his job and go to school. On the other hand, pursuing a college degree while maintaining full-time responsibilities for land-, sea-, and air-based tracking, and telemetry systems for NASA space flights seemed to Davis to be a daunting challenge and, as things turned out, it was. But through determination, perseverance, and self-discipline, Davis not only completed his college degree, but he also earned a promotion to Superintendent of Telemetry for RCA Corporation—a position that gave him responsibility for engineering functions relating to telemetry, timing, and firing systems from Cape Canaveral, Florida to South Africa.

(continued)

(*continued*)

Davis continued his climb up the career ladder, eventually becoming vice president of manufacturing for Harris Corporation. During his years with Harris Corporation, Davis earned a well-deserved reputation for making dramatic improvements in product reliability, customer satisfaction, employee involvement, efficiency, and profits by introducing various quality improvement initiatives. Over time, Davis established a reputation throughout Harris Corporation as a *quality guru*. In fact, he became so well known for quality improvement that when he retired from Harris Corporation, Davis was able to establish his own quality consulting firm, Stan Davis Consulting.

Davis now spends much of his time helping technology companies throughout the United States improve the quality of their people and processes as well as the value of their products and services. But Davis is more than a consultant. Like so many successful people, he is also a mentor. In addition to providing consulting services to technology companies that are trying to compete in a global business environment, Davis mentors executives in technology companies and helps prepare the executives of tomorrow by teaching college classes on quality-related subjects.

In order to spread his quality philosophy more broadly, Davis began to write books on various quality-related subjects. He is now a widely-read author on the subject of continual quality improvement. His books authored for Prentice Hall on quality management, ISO 9000, ISO 14000, and customer service are used throughout the United States as well as in many other industrialized nations throughout the world. One of his books on quality management has been translated into a foreign language (Indonesian).

He began his career as an electronics technician in the U.S. Air Force, but before retiring Davis had climbed all the way up the career ladder to the level of corporate executive. Then, at a point in life when most people kick back and take it easy, Davis launched a new career as a quality consultant, mentor, and author. His career exemplifies the concepts of self-discipline, mentoring others, and becoming a leader in your field as well as many of the other success strategies recommended in this book.

DISCUSSION QUESTIONS

1. Completing a college degree requires both self-discipline and good time management. Discuss how you have had to apply these concepts during your college years.
2. You have probably known college students who lacked self-discipline and time-management skills. Discuss how these shortcomings affected their college studies.
3. Discuss how the concept of *self-discipline* relates to the concept of *followership*.
4. Defend or refute the following statement: Self-discipline is one of those things you either have or you don't. Some people are born with it and some aren't.

ENDNOTE

1. John C. Maxwell, *Developing the Leader Within You* (Nashville, TN: Thomas Nelson, 1993), 161–162.

CHAPTER ELEVEN

Never Just Pass the Time—Use It

People count up the faults of those who keep them waiting.
   French Proverb

The most successful people use their time wisely. They have to because there are so many competing demands for their time. Professionals who fail to manage their time will often find themselves falling short on critical tasks as a result. Poor time management can cause a number of problems including the following: wasted time (yours and that of others), stress (on you and others), lost credibility, missed appointments, poor follow-through on commitments, poor attention to detail, ineffective execution, and poor stewardship of resources.

The following scenario shows how poor time management can affect performance. Janice was a department manager for a large engineering firm. She had excellent credentials and a strong background in her field, but Janice's team never seemed to live up to its potential. One of the reasons for this is that Janice was a notoriously poor time manager.

Because she was always running late, Janice never seemed to have time to spend with team members. As a result, she typically ignored their complaints, concerns, ideas, and suggestions. Her most frequent response to employees and colleagues was, "I'll get with you later. I don't have time right now." Of course, concerns and complaints that are ignored—whether from employees or colleagues—typically lead to problems that just get worse as time goes by. The longer they are ignored, the bigger the problems get. This happened to Janice all the time. As a result of her poor time management, Janice never got beyond the level of department manager in her company.

SUCCESS TIP

The most common causes of time-management problems are unexpected crises, telephone calls, no planning or poor planning, biting off more than you can chew, unscheduled visitors, reluctance to delegate, disorganization, lack of technology skills, refusing to say "no," and inefficient meetings.

CAUSES OF TIME-MANAGEMENT PROBLEMS AND WHAT TO DO ABOUT THEM

Here is a common scenario in today's workplace. A successful engineer comes to work with a list of what she hopes to accomplish that day only to have most of her time diverted to unplanned, unscheduled activities. The reasons this happens vary, but most of them are predictable. The most common causes of time-management problems are unexpected crises, telephone calls, no planning or poor planning, biting off more than you can chew, unscheduled visitors, reluctance to delegate, disorganization, lack of technology skills, refusing to say "no," and inefficient meetings. The most successful people have learned how to overcome these time robbers by applying good time-management strategies. Some effective time-management strategies you can use follow.

Unexpected Crises

Crises go with the job for most engineering professionals. It is a fact that the better you plan, the fewer crises you will face. However, even with good planning, crises will happen. Events over which we have no control can create unexpected circumstances that must be dealt with. Consequently, it is wise to assume there will be crises, and act accordingly. The following strategies can minimize the amount of time you have to devote to putting out fires.

- *Do not overbook.*   Anyone who has been to a doctor's office is familiar with the problem of overbooking. Typically, by mid-morning the doctor's schedule is already backed up. All it takes is one crisis with a patient, and the rest of the day is thrown off schedule. The solution to this situation is to schedule loosely. By scheduling loosely, you build in time to deal with the predictable occurrences of crises. The key to loosening your schedule is learning to actually schedule catch-up time between appointments.
- *Do not become an adoption agency for the unrelated problems of others.*   Successful people are can-do people who, when they see a problem, take the initiative to solve it. Normally this is precisely what you should do, but not always. Occasionally someone at work will bring you a problem that falls outside your range of responsibility or sphere of influence. Often such problems are personal in nature. In such cases, there is nothing wrong with offering advice or making a helpful referral. In fact,

SUCCESS TIP

Do not become an adoption agency for the personal problems of your direct reports or colleagues. Problems that relate to your job and areas of responsibility should receive your full attention. Unrelated problems, on the other hand, should not be allowed to monopolize your time.

providing this type of help is recommended. But advice and referral are as far as your help should go in such cases. Do not become an adoption agency for the personal problems of your direct reports or colleagues. Problems that relate to your job and areas of responsibility should receive your full attention. Unrelated problems, on the other hand, should not be allowed to monopolize your time.

Telephone Calls

Unless you manage its use wisely, the telephone can rob you of valuable time. Cellular telephones have only magnified the amount of time taken up by telephone calls. Fortunately, cellular telephones can also help you solve this problem. The following strategies will help minimize the amount of time in your day that is tied up by the telephone.

- *Increase your use of e-mail.* One of the best ways to avoid wasting time on hold, playing telephone tag, or listening to recorded messages is to use e-mail wherever possible. E-mail is not always a feasible option. However, when it is you can simply click on "Send" and move on to your next task; no pressing one for this option or two for that option, no talking to answering machines, and no waiting while on hold. In addition, people are often better about promptly returning e-mail messages than they are about returning telephone calls.

- *Separate important, routine, and unimportant calls.* Time invested in helping secretarial personnel learn to distinguish between important and unimportant telephone calls and between important and routine calls will be time well spent. One of the ways to do this is to provide secretarial personnel with a priority list of people you always want to talk to. In addition, train secretarial personnel to take comprehensive, detailed telephone messages. This will help you determine which calls should be returned and in what order.

- *Return telephone calls between meetings and during breaks.* On the one hand, cellular telephones can be obnoxiously intrusive. How many times have you been interrupted by the inopportune ringing of a cellular telephone? On the other hand, cellular telephones can help you turn time that might otherwise be wasted into productive time. You can save valuable time by taking telephone messages to meetings with you and using your cellular telephone to return them during breaks and between meetings. You can also use them to return calls from your car while traveling to your next meeting, provided you have the appropriate "hands-free" technology or that your car is off the road and parked. Don't be one of those people who drive with their knees while using a cellular telephone.

- *Block out call return times on your calendar.* Telephone tag is one of the most persistent time wasters in the workday. Assume, for whatever reason, that e-mail is not appropriate and you really need to talk to someone. You place the call, but the person you need to talk to is not available. You leave a message. This person wants to talk with you too, so she calls you back, but you are tied up in a meeting. She leaves a message. You call her back, but just miss her. This frustrating process repeats itself continually every day in businesses around the world. To minimize the amount of time you waste playing telephone tag, block out times on your calendar for returning calls and let the times be known to callers who leave messages. An effective approach is to schedule two thirty-minute blocks (at least) each day; one in mid-morning and one in mid-afternoon. Block these times out on your calendar as if they are appointments. Make sure that whomever takes your messages (even if it is a recording device) lets callers know that these are the times during which you typically return calls. In this way, if the caller really needs to talk with you, he will make a point of being available during one of these times. In fact, he can make connecting even easier by indicating which time he prefers.

■ *Get to the point.*   One of the reasons telephone calls are such time robbers is the human tendency to engage in idle chitchat. You can save a surprising amount of time on the telephone by simply getting to the point, and by tactfully nudging callers to do the same. There is certainly nothing wrong with a few appropriate comments on the latest ball game, movie, or items in the news, but the amount of time devoted to unrelated issues should be kept to a minimum. Stay focused, stay on task, and tactfully help callers do the same.

Poor Planning

It has already been mentioned that crises are a common time waster at work. There are crises that simply cannot be avoided, and there are crises that result from poor planning. A good rule of thumb to remember is this: The more effort you put into planning, the less effort you will waste dealing with crises. The following strategies can help improve your planning.

■ *End each day by planning the next.*   Devoting just ten to fifteen minutes at the end of each day to reviewing and planning for the next day can save valuable time. One of the best ways to minimize the occurrence of crises is to know when you walk in the door of your office every morning what is on your plate for the day. Three of the most persistent causes of crises are forgotten meetings, overlooked obligations, and missing paperwork. You can eliminate these causes altogether by ending each workday by reviewing the next day's meetings and obligations, retrieving any paperwork or other documentation that will be needed and familiarizing yourself with it, and creating a *to-do* list with the items listed in priority order. With just these few tasks accomplished, you will be able to begin the next day ready for what awaits you.

■ *Keep in mind that most tasks take longer to complete than you think they will.*   This is a good rule of thumb to follow. No matter what you have planned to do, experience shows it will probably take longer than you think it will. Consequently, it is wise to build a little extra time into your schedule. For example, if you think an appointment will take thirty minutes, allow forty-five minutes, and then try to finish in thirty. In this way, you have the extra time if it is needed without having to rush through the appointment. If the appointment happens to conclude on time, you can always put the extra time gained to good use returning telephone calls or getting a head start on other obligations.

SUCCESS TIP

Three of the most persistent causes of crises are forgotten meetings, overlooked obligations, and missing paperwork. You can eliminate these causes altogether by ending each workday by reviewing the next day's meetings and obligations, retrieving any paperwork or other documentation that will be needed and familiarizing yourself with it, and creating a *to-do* list with the items listed in priority order.

Trying to Do Too Much

Successful people take the initiative and seek responsibility. This is one of the reasons they are successful. This is also one of those ironic situations in which the good news is also the bad news. The good news is that successful people take the initiative, and the bad news is that successful people take the initiative. When you are a person who takes the initiative and seeks responsibility, it is easy to get out of balance and take on too much. When this happens, the following strategies can help you free up some valuable time.

- *Write down all current and pending tasks, projects, and obligations.*   Make a list of all of your current commitments. Then prioritize them. For each entry on your list, ask the following question: "What will happen if I don't do this?" This will usually give you a good start on paring down the list. Once you have eliminated any unnecessary obligations, prioritize the rest.
- *Examine all outside activities.*   The most successful engineering professionals are typically active in outside organizations such as civic clubs, chambers of commerce, economic development councils, and professional associations in their fields. Participation in these and other outside activities is an excellent way to grow as a leader and to benefit your organization. However, it is easy to fall into the trap of joining too many outside organizations or taking on too many responsibilities in these organizations. Balance is the key. For engineering professionals who want to succeed, outside activities are like food. Certain amounts of the right types are essential, but too much—even of the right types—can be harmful.

Unscheduled Visitors

One way to help regulate your time is to encourage people who want to meet with you to make an appointment. People who just drop in unannounced can rob you of valuable time. The following strategies can help minimize the amount of time in your day taken up by drop-in visitors.

- *Do not allow drop-in visitors during peak times.*   Some days are busier than others and some times of the day are busier than other times. During these peak times, it is best to ask drop-in visitors to come back at another time when you can give them

Success Tip

Poor delegation is one the easiest time-wasting traps to fall into. Often engineering professionals who become team leaders find it difficult to let go of work they are accustomed to doing themselves. In addition, some suffer from the *nobody-can-do-it-right-but-me* syndrome. These two phenomena can result in poor delegation, a major time waster.

your undivided attention—unless they are bringing you critical information or informing you of an emergency.

■ *Train secretarial personnel to rescue you.*   You can minimize the intrusions of drop-in visitors by working out an arrangement with secretarial personnel to rescue you after a certain amount of time (e.g., five minutes). It works like this. Whenever a drop-in visitor has been in your office for five minutes or so, the secretary buzzes you or looks in and says "it's time to place that important call," or "it's time for your next meeting or next appointment." This will tactfully let the drop-in visitor know that you have work to do.

■ *Remain standing.*   One way to convey the message that you are busy without having to actually say it is to remain standing when an unannounced visitor walks into your office. Once a visitor sits down and gets comfortable, it can be much more difficult to uproot him. By continuing to stand, you tactfully convey the message that "I can give you a few minutes, but only a few."

Poor Delegation

Poor delegation is one the easiest time-wasting traps to fall into. Often engineering professionals who become team leaders find it difficult to let go of work they are accustomed to doing themselves. In addition, some suffer from the *nobody-can-do-it-right-but-me* syndrome. These two phenomena can result in poor delegation, a major time waster. Tasks that do not require your level of expertise should be delegated. If subordinates cannot perform the tasks in question satisfactorily, you have a training problem, and training problems cannot be solved by refusing to delegate work.

Personal Disorganization

You can waste a lot of time rummaging through disorganized stacks of paperwork looking for the folder, form, or document needed. I once worked with an individual who had the unfortunate habit of never putting files, drawings, documents, or forms in the same place twice when he was done with them. Wherever this person happened to be when he finished with a file is where he would put it down. As a result, this otherwise talented engineer could be counted on to waste valuable time looking for "missing" paperwork. He eventually earned a reputation for being disorganized—a reputation that hurt his career. The following strategies can help minimize the amount of time you might waste as a result of personal disorganization.

- *Clean off your desk.* This strategy sounds so simple one might be tempted to discount it. But before doing so, look at your work area. Check your in-basket and your pending basket. Is there paperwork that is no longer relevant? Go through everything on your desk or in your work area and get rid of anything that is no longer pertinent in terms of your most pressing obligations. When trying to get organized, your best friend can be a large trash can.

- *Restack your work in priority order.* Go through your in-box or stack of pending work, and organize all work in order of priority. Work is often stacked in the order it comes in, especially when you are in a hurry and don't have time to organize it. Because this can happen so frequently, it's a good idea to occasionally stop what you are doing just long enough to go through your work stack and reorganize everything by priority. It's an even better idea to screen work as it comes in, placing your work in priority order at the outset.

- *Make use of categorized work files.* Teach clerical personnel to organize your paperwork by category. This means have a *Read Folder* for paperwork that should be read, but requires no writing or other action. Have a *Correspondence Folder* for nonelectronic correspondence you need to answer or initiate. Have a *Signature Folder* for paperwork that requires your signature (correspondence, requisitions, etc.). Organizing work in this way allows you to immediately locate what needs to be done without having to waste time sorting through stacks of paperwork.

Inefficient Use of Technology

Even well-educated engineering professionals are sometimes guilty of inefficient use of technology. Time-saving technologies save time only if you know how to use all of their various features, and use them well. The following strategies can help improve how efficiently and effectively you use time-saving technologies: (1) make sure you can use all of the various functions on your telephone (e.g., messaging, number storing, call waiting, park, camp, automatic redial, forward, save, repeat); (2) learn to dictate correspondence using a microcassette recorder; (3) make sure you can use all of the time-saving features on your computer; (4) make sure you can operate your Fax machine; (5) make sure you can use all of the time-saving functions on your copy

SUCCESS TIP

Much of the time spent in meetings is wasted. There are many reasons for this, the most prominent of which include poor preparation, the human need for social interaction, idle chitchat, interruptions, getting sidetracked on unrelated issues, no agenda, and no prior distribution of backup materials.

machine (sorting, collating, etc.); (6) equip your car as a second office; and (7) learn to use all of the features of your various handheld electronic devices.

Unnecessary and Inefficient Meetings

In spite of their value in bringing people together to convey information, brainstorm, plan, and discuss issues, meetings can be one of the biggest time wasters in the workplace. You can minimize the amount of time wasted in meetings by (1) meeting only when necessary; and (2) keeping necessary meetings as short as possible. Strategies that will help minimize the time wasted by meetings follow.

- *Understand what causes the wasted time associated with meetings.*   Much of the time spent in meetings is wasted. There are many reasons for this, the most prominent of which include poor preparation, the human need for social interaction, idle chitchat, interruptions, getting sidetracked on unrelated issues, no agenda, and no prior distribution of backup materials. In addition to these time wasters, there is also the *comfort factor.* Coffee, goodies, social interaction, and proximity to the boss can create such desirable environments that people simply don't want meetings to end. Make a point of eliminating these time wasters from meetings; especially those meetings you chair.

- *Examine all regularly scheduled meetings carefully.*   Most organizations have weekly, biweekly, and monthly meetings of various groups and teams. When these meetings were established they had a definite purpose, but over time, that purpose can become blurred or even go away. The meetings continue, though, if only out of habit. If you call or attend regularly scheduled meetings, ask the following questions about them: (1) Is the meeting really necessary? (2) What is the purpose of the meeting? (3) Could the meeting be scheduled less regularly (Can weekly meetings meet twice a month instead? Can monthly meetings meet quarterly instead? etc.)? (4) Could the purpose of the meeting be satisfied some other way (e-mail updates, written reports, etc.)?

- *Hold impromptu meetings standing up.*   Meetings that really should last no more than ten minutes can be kept on schedule by holding them standing up. These are typically meetings without an agenda called on an impromptu basis to quickly convey information to a select group or to get input from that group. Also, try to hold them in your office rather than in a conference room. It will be hard to keep participants from pulling up chairs and settling in if the meeting is held in a conference room.

- *Before meetings, complete the necessary preparations.*   For sit-down, scheduled meetings, have an agenda that contains the following information: purpose of the meeting, starting and ending time, list of agenda items with a person responsible for each, and a projected amount of time to be devoted to each agenda item. Set a deadline for submitting agenda items and stick to it. Require all backup material to be provided at the same time as the corresponding agenda items. Distribute the agenda, backup material, and the minutes of the last meeting at least a day before the meeting. If you distribute meeting materials too far in advance, participants will simply put them aside and forget about them. In addition, you increase the likelihood of cutting off the submittal of agenda items too soon. On the other hand, during the meeting if you wait until the materials get distributed you will waste time handing them out and waiting while participants read them. Ask all participants to read the agenda and backup materials before the meeting. This is why it is distributed beforehand in the first place.

■ *During the meeting stay focused and stick to the agenda.* Begin meetings on time. Waiting for latecomers only encourages tardiness. If participants know you are going to start on time, most will eventually discipline themselves to arrive on time. Have someone take minutes. In the minutes, all action and follow-up items should be typed in bold face so they stand out from the routine material. Make the minutes of the last meeting the first item on the agenda. In this way the first action taken in the meeting is following up on assignments and commitments made during the last meeting. Stay focused. Keep participants on the agenda and on task. The last agenda item should always be either "New Business" or "Comments From the Floor." Such an agenda item gives participants an opportunity to bring up issues that are not on the agenda without getting the meeting sidetracked before agenda items have been disposed of. Let participants know that only critical issues and bona fide emergency cases are to be brought up as new business. Otherwise, participants will begin to bring up all of their issues in the new business portion of the meeting rather than devoting the preparation time necessary to get them on the agenda. Ask participants to turn off cellular telephones during meetings. Interruptions from cellular telephones can be a major distraction and time waster in meetings.

■ *After meetings, follow up quickly.* Have the minutes of the meeting typed and distributed right away—ideally on the same day the meeting occurred. E-mail distribution simplifies this task. Allow an appropriate amount of time for participants to act, then follow up on action items from the minutes. If you call meetings, never wait until the next meeting to ask about progress made toward completing the action items from the previous meetings. Ask periodically between meetings.

The strategies recommended in this chapter will help you take control of your most valuable asset—time. If you put these strategies to good use, your performance and, in turn, your career will be enhanced.

REVIEW QUESTIONS

1. Explain how you can minimize the amount of time you have to spend on unexpected crises.
2. Explain how you can minimize the amount of time in your day devoted to telephone use.
3. List and explain several strategies for improving your planning.
4. If you are one of those people who try to do too much, explain some strategies you can use to free up valuable time.
5. How can you minimize the amount of time that will be taken up by unscheduled visitors?
6. Explain how proper delegation can give you more time.

7. List and explain several strategies for personal organization.
8. List and explain several strategies for using technological devices as time savers.
9. Explain how to minimize the amount of time wasted in meetings.

DISCUSSION QUESTIONS

1. Now that cellular telephones are so readily available, you never have to be out of touch. This is the positive aspect of cellular technology. The negative aspect is that cellular telephones can ring anytime and be obnoxiously intrusive. Discuss ways to use a cellular telephone as a time-saving device while keeping it from becoming a time-robbing device.

2. Assume you are a team leader who prides yourself on having an open door for your team members. Unfortunately, some of your team members take advantage of your open door to tie up a lot of your time telling you about their personal problems. Discuss ways you can minimize the amount of time spent listening to the personal problems of your team members.

3. Your boss is the most disorganized person you have ever met. He never plans, is habitually late, and is constantly looking for files and other materials he needs but has misplaced. Discuss the problems this might cause in your department and strategies you might recommend for getting the boss organized.

CHAPTER TWELVE

Learn To Be Customer Driven

Every company's greatest assets are its customers, because without customers there is no company.[1]

Michael Le Boeuf

In today's business world the customer is the king. In order to survive and thrive in today's global business arena, engineering firms must be able to outperform the competition. The judge who decides which firm wins the daily competition is the customer. Consequently, an important characteristic shared by engineering professionals in today's workplace is a customer-driven perspective. Engineering professionals with a customer-driven approach to their jobs keep customer satisfaction in the forefront of their thinking and their actions. Once you complete college and begin your career, ensuring customer satisfaction will contribute greatly to your career advancement.

While pursuing his engineering degree, Mike worked as a CAD technician for an engineering services firm. His company's customers were architects who designed commercial buildings. Mike's company developed the civil, structural, mechanical, and electrical engineering plans that, along with the architectural drawings, became the plans for the buildings. It was not uncommon for architects to send incomplete information to Mike's company. It was this architectural information that formed the basis of the various design calculations and decisions that would be made by the engineers at Mike's company.

Most of the engineers and experienced CAD technicians at Mike's company refused to begin an architect's project until all information provided was complete. This approach was understandable from the engineer's point of view, but caused much consternation on the part of the architects—the customers—who could not afford to have their projects fall behind schedule. Unlike his colleagues, Mike understood that incomplete information, though inconvenient, was just part of the job.

Consequently, rather than refuse to begin a project until all information provided by the architects was complete, Mike would find ways to work around the missing information until it could be provided or cleared up. He would get as much done on a project as could be safely done while simultaneously working the telephone to obtain the missing information. Sometimes by studying the architect's drawings and comparing the information provided in them with other architectural materials such as specification documents, Mike could find his own answers and fill in the missing information himself.

As a result of his helpful approach, architects began to request Mike to work on their projects. As this happened more and more frequently, the CEO of Mike's company began to ask customers—the architects—why they preferred Mike to others in his company. What he learned probably saved his company a good deal of future business that might have migrated to the competition as architects became frustrated with the engineers and technicians in his company who refused to budge until they had complete information.

The CEO was so impressed with what he learned about Mike that he not only offered him a full-time engineering position upon graduation from college, but also agreed to pay Mike's tuition and book costs if he would accept the position. Mike completed his degree on schedule and was promoted not to the traditional engineer-in-training position, but directly to a project engineer status. His experience as a CAD technician coupled with his philosophy of excellent customer service launched Mike's career with this company; a career that eventually saw him serve as the company's chief engineer.

WHAT IS A CUSTOMER?

Customers go by many different names depending on the type of organization in question. Healthcare organizations call their customers *patients*. Technical consulting firms call their customers *clients*. Certain types of publishing firms call their customers *subscribers*. Other organizations call their customers just that—*customers*. Regardless of title, *a customer is a person or an organization with whom you exchange something of value.*

Engineering firms provide services of value to their customers and, in return, their customers provide something of value to them—money. Manufacturing, processing, and other technology companies provide products of value to their customers and, in return, their customers provide something of value to them—again it's money. This mutually beneficial exchange of value is the heart and soul of business, regardless of the type of business in question.

When the term *customer* is used, most people are referring to those people and organizations that use their products or services—external customers. External customers will always be your most critical customers. However, there are also *internal customers*. An internal customer is simply one employee whose work is done to assist another employee. For example, when computer-aided drafting technicians develop drawings for engineers, the engineers are their internal customers. Successful engineering professionals understand the concepts of external and internal customers and use this understanding to guide their actions and decisions.

SUCCESS TIP

Being a customer-driven professional is so important because customer-driven behavior is the best way to ensure customer satisfaction, and customer satisfaction is essential to success in business—yours and your company's.

When you first begin your career in engineering, you will probably focus more intently on internal customers. As you are promoted to higher positions with increased levels of responsibility, your customer focus will be more and more external. As this happens, don't forget that internal customers are important too.

WHY BEING CUSTOMER DRIVEN IS SO IMPORTANT

When you complete your college studies and begin your career, few things will be so important to your career development as being a customer-driven professional. Customer satisfaction is critical in a competitive marketplace. If customers dislike your organization's products or services or if customers are unhappy about how your organization treats them, they can take their business elsewhere. Every time this happens, your organization loses and some other organization wins. Being a customer-driven professional is so important because customer-driven behavior is the best way to ensure customer satisfaction, and customer satisfaction is essential to success in business— yours and your company's. Correspondingly, customer dissatisfaction is the surest way to guarantee failure.

Consider the following facts that are widely known in business circles about dissatisfied customers:

■ Dissatisfied customers don't just take their business elsewhere, they tell other potential customers about their dissatisfaction with your company.

■ Few dissatisfied customers will bring their issues to you and give your organization a chance to correct their problem. Instead, most will simply take their business elsewhere.

■ Once dissatisfied customers migrate to a competitor, it is very difficult—some believe impossible—to get them back.

■ It typically costs an organization ten times more to attract new customers than to retain existing customers.

From these few facts you can see that being a customer-driven professional is good business and good career advancement advice.

STRATEGIES FOR BEING CUSTOMER DRIVEN

Being a customer-driven professional is about more than just attitude and outlook. It's also about action. What follows are a number of specific strategies you can apply to transform your customer-driven attitude into action.

Take Responsibility for Customer Problems and Complaints

Consider the following scenario. The telephone in your office rings. You answer it. The person calling is a customer with a problem, but the problem is not in your area of responsibility. This happens frequently in the workplace. When it does, you have a couple of options. The first option is what I call the *handoff*. With this option you might

SUCCESS TIP

> With technical products, customers are often the cause of their own problems. Whoever said the customer is always right never developed software packages. However, although the customers may not be right when they complain, they should always be treated right.

say, "Your problem is not in my department. You need to talk to someone from the mechanical engineering department. That extension is 320." The handoff option is bad business. It gives customers the impression you don't want to be bothered with them.

The other option is the customer-friendly approach. With this option you take responsibility for the customer and make the connection for him with the right person. Rather than just telling him to hang up and call another number, you might say, "Thanks for calling. I'll connect you with someone who can help you right away." Then you transfer the call to the right person, making sure that person answers the telephone before you disconnect from the customer.

If the person who can solve the problem is not readily available, you ask the customer to relate the problem to you. Then you track down the person who can solve the customer's problem and have that individual call the customer back as soon as possible. Before disconnecting from the customer, you should let him know what you are going to do. This customer-friendly approach will make a much more positive impression on the customer than transferring him to voice mail or telling him to hang up and call another number.

Find Permanent Solutions—Not Temporary Fixes

Engineering professionals are always busy. Consequently, when customers make complaints or point out problems, the temptation is to arrange a quick temporary fix. The problem with quick temporary expedients is that the original problem still exists. The better approach is to work with customers to find permanent solutions to their problems. Permanent solutions might take longer to put in place, but they are *permanent*. Once applied they stay applied. Dealing with a string of temporary fixes will take more time in the long run, and will just make the customer increasingly unhappy.

Help Customers Even When the Problem Is Their Fault

With technical products, customers are often the cause of their own problems. Whoever said the customer is always right never developed software packages. However, although the customers may not be right when they complain, they should always be treated right. They might not have read the instructions correctly and, as a result, created a problem. But they are still customers and your organization wants them to continue to be customers—satisfied customers. This means you should help customers even when they are the source of their own grief. If you don't, a competitor will.

Success Tip

> Customers can be demanding. Because of this, customer-oriented engineering professionals trying to meet their demands sometimes make the mistake of promising more than they can deliver. When working with customers, it's better to promise small and deliver big than to make a promise you cannot keep.

Value the Customer's Time

When dealing with customers, remember that their time is valuable too. Don't make them wait, and when working with them, be efficient. Organization, planning, and preparation are the keys to dealing with customers in a time-sensitive manner. Before meeting with customers, do your homework, organize your paperwork, and plan your agenda. Customers who have their time wasted by your organization will sooner or later migrate to a competitor that respects their time.

Greet Customers with Enthusiasm

Customers should never be viewed as an inconvenience or an intrusion. Customers are why your organization is in business, so treat them accordingly. Welcome their telephone calls and office visits with enthusiasm. Customers can sense when they are appreciated. The minute customers sense otherwise, they will begin looking for a competitor that will appreciate them.

Look for Customer Irritants

Often it's the little things that bother customers most. When customers call your company, do they get tied up forever in the telephone answering system? Does your company have a welcome "feel" to it? Does your company facilitate the process to make it convenient when customers need to visit restricted-access areas? Does your company have a comfortable place for customers to wait when they show up early for meetings? Are the customers' preferences considered when setting up business luncheons? Some of the issues relating to these questions might seem trivial to you, but it's often the little things that make the biggest difference with customers and in advancing your career.

Promise Small—Deliver Big

Customers can be demanding. Because of this, customer-oriented engineering professionals trying to meet their demands sometimes make the mistake of promising more than they can deliver. When working with customers, it's better to promise small and deliver big than to make promises you cannot keep. For example, say you are talking on the telephone with a customer who is pressuring you to get a set of engineering

SUCCESS TIP

> Once you have reached a high-enough level in your organization to be heard on such matters, begin encouraging the involvement of customers in all aspects of the organization. The point of this exercise is *continual improvement*—an essential concept in a competitive environment.

drawings delivered to him as soon as possible. You think that by using an overnight express service you can probably get the drawings there by noon tomorrow, but you're not sure.

Rather than set yourself up for a customer complaint about a late-arriving package, promise small. Tell the customer you can get the package to him by 4:00 pm instead of noon. Then do everything you can to get the package to him by noon. In this way, when the package arrives by noon or before, you are the hero not the goat. Not only was the package on time, but it was actually several hours early. This is called promising small but delivering big, and it's an effective customer-service strategy.

Let Customers Know About Changes Before They Are Made

If your organization makes any kind of change that will affect customers in even the slightest way, let them know about it *before* the change is implemented. Customers don't like surprises. They become accustomed to your organization's procedures and plan their dealings with you accordingly. For example, if your e-mail address, telephone number, or mailing address is going to change, let customers know well before it happens.

If the engineers your customers are accustomed to dealing with change, let customers know before it happens. If you change offices, let customers know how to find your new location. Any time any type of change is contemplated, before making the change, ask yourself, "How will this change affect customers?" Once you have answered this question, take whatever action is necessary to mitigate any negative consequences for customers.

SUCCESS TIP

> Many organizations ask for customer feedback—an after-the-fact approach to involving customers. But the most competitive organizations ask for customer input. In other words, they involve customers *before* introducing or changing a product or service. In this way, these organizations know beforehand how customers are going to react. If the reaction is positive, the new idea goes forward. If the reaction is negative, the new idea is either revised or replaced with a better idea, and customers help you decide what the revision or the new idea should be.

Involve Customers in Your Organization

Once you have reached a high-enough level in your organization to be heard on such matters, begin encouraging the involvement of customers in all aspects of the organization. The point of this exercise is *continual improvement*—an essential concept in a competitive environment. The judge who will ultimately determine the value of the products and services your organization provides is the customer. Consequently, as you work to continually enhance that value, it makes sense to involve customers.

Many organizations ask for customer feedback—an after-the-fact approach to involving customers. But the most competitive organizations ask for customer input. In other words, they involve customers *before* introducing or changing a product or service. In this way, these organizations know beforehand how customers are going to react. If the reaction is positive, the new idea goes forward. If the reaction is negative, the new idea is either revised or replaced with a better idea, and customers help you decide what the revision or the new idea should be.

The most competitive organizations solicit customer input through a variety of means. Some use customer focus groups in which teams of customers are formed and asked to give their input concerning a product or service that is under development. Some use customer surveys. Others select specific loyal customers whom they trust and then meet with them one-on-one. The actual method used is less important than the fact that these competitive companies involve customers at every point in the developmental process. After all, who knows better what the customer will like than the customer?

Establish a Customer-Focused Infrastructure in Your Organization

You will soon find after graduating from college and beginning your engineering career that there are more organizations that talk the talk than walk the walk of customer service. You can help advance your career by helping your company walk the walk of customer service. One of the best ways to do this is to ensure that your organization establishes a customer-focused infrastructure.

The missing ingredient in many organizations that talk a lot about effective customer service but never seem to actually provide it is a supportive foundation. Remember this in all you do during your career. People in organizations typically do what they are (1) expected to do, (2) held accountable for doing, and (3) rewarded for doing. This means that if you want to have effective customer service in an organization, you need to expect, monitor, evaluate, recognize, and reward it. The means by which you do these things is what I call your organization's *performance infrastructure*. It has the following components: strategic plan, job descriptions, performance appraisal system, daily monitoring by supervisors, and the recognition and reward system.

If effective customer service is important, the organization's strategic plan should say so. By this I mean there should be at least one written corporate value or guiding principle in the strategic plan stating that effective customer service is a high priority in the organization. This will show that executive management *expects* effective customer service. If effective customer service is part of every employee's job—as it should be—then every employee's job description should say so. This will clearly show all employees that customer service is an important part of what they are *expected* to do every day.

If effective customer service is included in the organization's strategic plan and job descriptions, all management and professional personnel should monitor the customer-service behaviors of employers daily. What managers and other professionals monitor will show employees what they can expect to be held *accountable* for. If a certain type of behavior is expected and monitored, it should also be periodically evaluated. This means that customer service should be included as a criterion in the organization's performance appraisal system. If employees know their customer-service-related behavior will be evaluated periodically, they know they are being held *accountable* for customer service.

If employees are expected to demonstrate positive customer-service behaviors and they are accountable for doing so, they should be properly *recognized* for effective performance. This means that when employee recognition awards (e.g., employee of the month, quarter, year) are given, customer-service behaviors should be one of the selection criteria. Finally, if positive customer-service behaviors are important, the organization should *reward* them. This means that when decisions are made about promotions, salary increases, and monetary bonuses, customer-service behaviors should be an important factor in making the decisions.

In working to advance your career, it will be important for you to apply a customer-positive attitude from the outset, and to continue doing so throughout your career. As your career develops and you achieve higher and more responsible positions, it will be important that you make sure your organization is customer friendly. If you begin applying the concept of *customer involvement* as soon as you secure your first professional position after college, your career will benefit noticeably.

A complaint I hear frequently from engineering professionals is that recent college graduates don't seem to "get it" when it comes to customers. Too many recent graduates want to tell customers what they—the customers—should want rather than listening to determine what they really need. While this might be bad news for organizations trying to thrive in a competitive environment, it can be good news for you, provided the customer-involvement strategies presented herein become part of your daily routine and outlook.

REVIEW QUESTIONS

1. Use your own words to define the term *customer.*
2. Explain why it will be so important for you to be a customer-driven professional once you begin your career.
3. Explain what is meant by taking responsibility for customer problems and complaints.
4. Why is it important to find permanent solutions for customers rather than just quick temporary fixes?
5. Explain why you should help customers solve their problems even when they themselves created the problems.
6. Why is it important to value the customer's time?
7. Explain the importance of greeting customers with enthusiasm.
8. Why is it important to identify and eliminate minor customer irritants?

9. Explain the concept of promising small, but delivering big.
10. Why is it important to let customers know about changes before they occur?
11. What are some ways you will be able to involve customers in your organization once you finish college and begin your career?
12. How can organizations establish a customer-focused infrastructure?

DISCUSSION QUESTIONS

1. Discuss how you might handle the following situation once you are established in your career. A customer telephones your company and you answer the telephone. The customer is angry and wants help solving a problem. The problem has nothing to do with your department.
2. Defend or refute the following statement: If a customer can't read the instructions and causes a malfunction in the product, that's his problem.
3. Think of some business, store, restaurant, or other organization that you frequent. Are there any minor irritants that bother you about this organization? Discuss the irritants and what the organization could do to eliminate them.
4. Have you ever dealt with a business that promised something, but failed to deliver? Discuss how that made you feel about continuing to do business with that organization.

ENDNOTE

1. Louis E. Boone, *Quotable Business*, 2nd ed. (New York: Random House, Inc., 1999).

CHAPTER THIRTEEN

Develop and Apply a Can-Do Attitude and Seek Responsibility

People can overcome even the most difficult obstacles and go on to achieve great success if they have a positive, can-do attitude.

Anonymous

Winston Churchill was prime minister of Great Britain for the first time during the difficult years of World War II. He is viewed by students of history as the man who single-handedly held his country together during those early dark days of World War II when Great Britain stood virtually alone against Hitler's Nazi juggernaut. More than anything else, it was Churchill's *can-do* attitude that bolstered the resolve of his beleaguered countrymen.

The positive effect of Churchill's can-do attitude can be seen in the comments of his contemporaries:

> We owed a good deal in those early days to the courage and inspiration of Winston Churchill who, undaunted by difficulties and losses, set an infectious example to those of his colleagues who had given less thought than he, if indeed any thought at all, to war problems . . . His stout attitude did something to hearten his colleagues.[1]

From this quotation by one of Churchill's contemporaries we can see that at a time when his countrymen needed it most, Churchill displayed a can-do attitude and

SUCCESS TIP

A can-do attitude is the outward manifestation of an inner conviction that says whatever the job, you can get it done; whatever the challenge, you can meet it; and whatever the obstacle, you can overcome it. Such an attitude, although certainly optimistic, should not be confused with false bravado.

124

took responsibility for leading the defense of Great Britain. His is an excellent example for professionals who hope to lead teams, departments, or organizations in the battle of the marketplace and you win at building a successful career. The can-do attitude Churchill used to help his country eventually win in a global war is the same attitude you need to help your company win in the global marketplace.

WHAT IS A CAN-DO ATTITUDE?

A can-do attitude is the outward manifestation of an inner conviction that says whatever the job, you can get it done; whatever the challenge, you can meet it; and whatever the obstacle, you can overcome it. Such an attitude, although certainly optimistic, should not be confused with false bravado. Rather, a can-do attitude is founded on the conviction that, within the bounds of legality and ethics, you can and will do whatever is necessary to succeed. A can-do attitude has the following elements: optimism, initiative, determination, responsibility, and accountability.

The story of Herb Kelleher and Southwest Airlines demonstrates the value of a can-do attitude. Southwest Airlines is certainly not an engineering company, but Kelleher's example applies not only to engineering but to any other profession. Kelleher, an attorney by training, is one of the three founders of Southwest Airlines. One of the most successful airlines in the United States, Southwest almost never got off the ground. Before the ink was dry on its incorporation papers, Southwest Airlines found itself ganged up on by competing airlines that tried to keep it from ever getting off the ground with a series of expensive lawsuits. "One court battle followed another, and one man, more than any other, made the fight his own: Herb Kelleher. When their start-up capital was gone, and they seemed to be defeated, the board wanted to give up. However, Kelleher said, 'Let's go one more round with them. I will continue to represent the company in court, and I'll postpone any legal fees and pay every cent of the court costs out of my own pocket.' Finally, when their case made it to the Texas Supreme Court, they won, and they were at last able to put their planes in the air."[2]

The can-do attitude of Herb Kelleher helped him lead Southwest Airlines through many more struggles until the company eventually grew from a firm with four airplanes and total assets of $22 million to a successful airline with almost 275 airplanes and assets exceeding $4 billion.[3] Kelleher's example shows the value of a can-do attitude.

SUCCESS TIP

There is a tendency among some to think that an optimistic, can-do attitude is a gift people either have or don't have. Such people see a can-do attitude as something you are born with or without. Nothing could be further from the truth.

WHY IS A CAN-DO ATTITUDE IMPORTANT?

Part of succeeding in engineering involves becoming a leader in your organization. This is because the attitudes of leaders are often mirrored by their followers. A leader with a bad attitude is likely to spawn bad attitudes in his followers; if, that is, he can manage to maintain any followers. People do not like to follow someone with a bad attitude. On the other hand, as the earlier example of Winston Churchill clearly shows, a leader with an optimistic, can-do attitude can sustain the morale of his followers through even the most difficult of times. "Optimism is also the key to the can-do spirit, to the don't-take-no-for-an-answer attitude that is essential to successful executive leadership. Nearly all human organizations are subject to an inertia that results in an it-can't-be-done attitude. This was always unacceptable to Churchill."[4] Such an attitude should also be unacceptable to engineering professionals who want to lead.

YOU ARE RESPONSIBLE FOR YOUR ATTITUDE

There is a tendency among some to think that an optimistic, can-do attitude is a gift people either have or don't have. Such people see a can-do attitude as something you are born with or without. Nothing could be further from the truth. At the beginning of this chapter, Winston Churchill was used as an example of a leader who exemplified the can-do attitude. Churchill is such a good example precisely because he had to work so hard to maintain the can-do attitude for which he is so well known. What is less known, except by careful students of history, is that Churchill suffered from frequent and severe bouts of depression. His well-known can-do attitude was a hard-won attribute he had to strive constantly to maintain. It is the same with most people.

A can-do attitude is as much a tool to an engineering professional as a hammer is to a carpenter, and just as the carpenter is responsible for maintaining his tools, the professional is responsible for maintaining his attitude. The following quotation is often used to remind people that a positive, can-do attitude is a choice rather than a gift or an accident of birth:

> We cannot choose how many years we will live, but we can choose how much life those years will have. We cannot control the beauty of our face, but we can control the expression on it. We cannot control life's difficult moments, but we can choose to make life less difficult. We cannot control the negative atmosphere of the world, but we can control the atmosphere of our minds. Too often, we try to choose to control things we cannot. Too seldom, we choose to control what we can . . . our attitude.[5]

YOUR ATTITUDE AFFECTS THE ATTITUDES OF OTHERS

Consider the words of John C. Maxwell: "People catch our attitudes just like they catch our colds—by getting close to us. It is important that I possess a great attitude, not only for my own success, but also for the benefit of others."[6]

The fact that people will look to their leaders for the type of attitude they should adopt is precisely why it is so important for you to maintain a positive attitude.

SUCCESS TIP

When one person in an office gets a cold, it seems that soon everybody has one. The same can be said of a bad attitude. Further, just as it is easier to catch a cold than to prevent one, it is easier to spread a bad attitude than to prevent one.

In the above quotation, Maxwell makes the point that an attitude spreads in the same way a cold spreads.

This is an apt description of what actually occurs. When one person in an office gets a cold, it seems that soon everybody has one. The same can be said of a bad attitude. Further, just as it is easier to catch a cold than to prevent one, it is easier to spread a bad attitude than to prevent one. This is why it is so important for you to be both persistent and consistent in displaying a positive can-do attitude.

HOW TO DEVELOP A CAN-DO ATTITUDE

Some people may be born with a can-do attitude, but most are not. Developing a positive attitude is like developing a muscle: (1) it takes hard work, determination, and persistence; and (2) if you stop working at it, you can quickly lose it. This section describes a process for developing a positive can-do attitude.

Step 1: Assess

Self-assessment is always difficult and sometimes painful. We don't like to focus too much attention on our personal shortcomings. However, the ability to look at yourself objectively, identify problems, and do what is necessary to correct them is a mark of a mature professional who has the potential to succeed. When assessing your attitude, objectively identify any aspects of your thoughts, feelings, and behavior that need to be improved.

The results of the self-assessment conducted in this step are the beginning point for the plan you will develop in the next step. Use the following questions to conduct the self-assessment. The desired answer to each question is "Yes."

- Are your thoughts about people generally positive?
- Are your thoughts about your work generally positive?
- Can you typically disagree with people without being disagreeable?
- Do you typically maintain your composure when under pressure?
- Do you typically perform well under stress?
- In most situations do you feel like you can get the job done?
- Are you able to delegate work to people who do things differently than you?
- Do you typically seek responsibility instead of waiting for it to be assigned?

- Are you typically positive about finding solutions when unexpected problems arise?
- Does your behavior encourage perseverance when obstacles stand in the way of success?

Step 2: Plan

Your plan should have the following elements: improvement goals based on the self-assessment, specific actions to be taken to achieve the goals, and metrics for measuring progress.

- *Goals.*   Convert any weakness identified during the self-assessment into goals. State the goals in behavioral (doing or action) terms as an improvement you would like to make. For example, assume that you answered "No" to the following self-assessment question:

  Can you typically disagree with people without being disagreeable?

  This problem area could be converted into a behaviorally stated goal that would read as follows:

  Learn to disagree with people without being disagreeable.

  By converting the attitudinal problem into a behaviorally stated goal, you encourage improvement in two ways—both positive. First, you eliminate all ambiguity. Having written down the goal, there is now no question of what you need to do in order to improve. Second, you build in personal accountability by creating an expectation—in writing—so that progress can be measured.
- *Action Steps.*   Identify specific action steps you can take to accomplish each improvement goal you set. Action steps identify more specifically than the goal exactly what must be done in order to improve. For example, the following action steps would help achieve the goal of learning to disagree with people without being disagreeable.

  a. When I disagree with someone and feel myself getting angry, I will silently count until the anger impulse passes before speaking.
  b. I will make a list of tactful ways to disagree with people and put the list to use.

SUCCESS TIP

By converting the attitudinal problem into a behaviorally stated goal, you encourage improvement in two ways—both positive. First, you eliminate all ambiguity. Having written down the goal, there is now no question of what you need to do in order to improve. Second, you build in personal accountability by creating an expectation—in writing—so that progress can be measured.

■ *Metrics for Measuring Progress.*   There is a management principle that says, ". . . if you want to make progress, measure it." Measurement is a necessary factor in the formula for accountability. This is why people who are trying to lose weight are required to weigh themselves daily. If your action steps are stated in behavioral terms, they can be measured. This is important because human nature being what it is, you will make better progress if you invest time and effort to measure results. For example, progress in implementing the action step from the previous paragraph can be measured. How many times this week did you silently count before responding to someone with whom you disagreed? Is that more times than last week?

You might question the need to develop a *written* plan for improving your can-do attitude. People I have worked with often say, "Now that I've completed the self-assessment, I know what needs to be done. I don't need to write it down." However, experience has shown that those who apply this rationale rarely succeed. Those who invest the time and effort to develop a written plan are also more likely to invest the time and effort necessary to make improvements.

Step 3: Implement, Monitor, and Adjust

Once your plan is complete, implement the plan and use the metrics included in it to measure progress. If you are making acceptable progress, continue on course. If not, make adjustments. A plan is just that—a plan. If it's working, stay with it. If not, drop the ineffective strategies and try new ones. An old saying credited to the Apache Indians applies here: *If the horse you are riding dies, climb off and get on another.* In other words, don't become wedded to your plan. If the plan isn't working, revise it. It is the progress toward improvement that matters, not the specifics of the plan.

REVIEW QUESTIONS

1. Explain in your own words the concept of the *can-do attitude.*
2. Why is it important for those who want to build a winning career to have a can-do attitude?
3. Is a can-do attitude something you are born with or something you can develop? Explain.
4. How can your attitude affect that of others?
5. Explain the steps for developing a can-do attitude.

DISCUSSION QUESTIONS

1. Defend or refute the following statement: It's easy for you to say that I should have a can-do attitude when you were born with one but I wasn't.
2. Defend or refute the following statement: I wish Mary had a better attitude, but it's not her fault. She can't help having a bad attitude.

3. Have you ever known someone who had a bad attitude? If so, discuss how it felt to be around this person. Did you find that this person's bad attitude affected your attitude?

4. Discuss how you would answer the following question from one of your team members: I want to develop a can-do attitude, but don't know how. Can you help me?

ENDNOTES

1. Maurice Hankey as quoted in Steven F. Hayward, *Churchill on Leadership* (Rocklin, CA: Forum, 1998), 115.
2. John C. Maxwell, *Developing the Leader Within You* (Nashville, TN: Thomas Nelson, Inc., 1993), 163.
3. Ibid.
4. Hayward, *Churchill on Leadership*, 116.
5. Anonymous, "Attitude," *Barlett's Familiar Quotations*, ed. Emily Morison Beck (Boston: Little Brown, 1980), 413.
6. Maxwell, *Developing the Leader*, 105.

CHAPTER FOURTEEN

Don't Just Plan—Execute

Having a good plan is important, but effectively executing that plan is even more important.

Anonymous

Every year, two teams from the National Football League meet in the Super Bowl. On game day, fans on every continent jam into their favorite sports bars and pack each other's homes to cheer for their favorite team. Knowing that an international spotlight is shining on them, both teams come to this annual event well prepared—each with its own plan for winning the game. However, when the final whistle blows only one of the teams will take home the coveted Lombardi Trophy. With rare exceptions, the winner of the Super Bowl is the team that does the best job of executing its game plan. The same can be said of companies operating in a competitive business arena. Having a plan is important, but executing the plan is essential. You can use this fact to help advance your career. All you have to do is learn how to go beyond just planning to actually executing plans.

The best plan in the world is just a summary of good ideas until it is effectively executed. It is one thing to plan; it is quite another to execute. This chapter provides strategies you can use to earn a reputation for effective execution.

WHAT IS MEANT BY EFFECTIVE EXECUTION?

Effective execution is defined as follows: *Systematically initiating, following through on, and completing all tasks necessary to effectively implement a plan. Execution involves applying such strategies as establishing expectations, assigning responsibility (delegation), establishing accountability, allocating resources, identifying and overcoming inhibitors, monitoring progress, flexibly adapting as necessary, and following through.*

There is a saying that the devil is in the details. This saying certainly applies when executing a plan. Whereas planning is about conceptualizing and envisioning the future, executing is a practical, hands-on enterprise that deals with the details of the here and now. The most successful engineering professionals are those who have learned to successfully handle the details of execution.

SUCCESS TIP

Developing a good game plan and executing it are two very different challenges. Plans are developed in the air-conditioned comfort of offices and conference rooms under conditions that are easily controlled. Plans are executed in the arena where your competitor is trying not only to execute its plan, but also to prevent you from executing yours.

WHY GOOD PLANS OFTEN FAIL

This chapter opened with an analogy in which both football teams came to the Super Bowl with solid plans for winning. Neither team will execute its game plan perfectly, but provided they both have good plans, the one that comes closest to an effective execution is the one most likely to win. Rarely is the Super Bowl won just because Team A had a better game plan than Team B. The teams that play in the Super Bowl know each other. The coaches on both teams are well versed in the comparative strengths and weaknesses of the teams and their individual players. Consequently, both coaching staffs are able to develop solid game plans that, if effectively executed, will ensure victory.

Developing a good game plan and executing it are two very different challenges. Plans are developed in the air-conditioned comfort of offices and conference rooms under conditions that are easily controlled. Plans are executed in the arena where your competitor is trying not only to execute its own plan, but also to prevent you from executing yours. To illustrate this point, I return to the Super Bowl analogy.

Assume that a strategy in Team A's game plan is to run the ball from scrimmage fifty times during the game, and gain 250 rushing yards. This strategy is built into the game plan because Team A's coaches know from researching Team B that the only teams that have managed to beat Team B in the last three years were those that rushed for 250 yards or more. Consequently, this is a well-founded strategy. However, it is easier to adopt this strategy than it will be to execute it. During the game, every time Team A's quarterback calls a running play, there will be eleven players on Team B doing their best to make sure the play fails.

In order to gain 250 rushing yards in fifty carries, Team A will need to average five yards per carry. In order to do this, every player on Team A will have to effectively execute his assignment on every rushing play—a difficult challenge under even ideal circumstances. But circumstances on the football field and in the workplace are rarely, if ever, ideal.

SUCCESS TIP

Another factor that can undermine the effective execution of plans is what I call the *dirty hands syndrome*. This syndrome is a reluctance on the part of people to step outside of the safe and comfortable intellectualism of planning, and get their hands dirty sorting through the often-messy details of execution.

Success Tip

Planning involves considering the possible, envisioning the ideal, and looking to the future. It is an appealing intellectual activity full of promise and hope. Execution, on the other hand, involves digging into the details, dealing with reality, and focusing on the here and now. It is a practical, roll-up-your-sleeves-and-get-your-hands-dirty activity that is tempered by the practical realities of business and life.

For example, consider Team B's game plan. Team B's coaches know that Team A typically depends more on its running game than on its passing game. Consequently, a strategy in Team B's game plan is to hold Team A to less than one hundred rushing yards. This means that every time Team A and Team B line up across from each other, Team B's defense will be trying to undermine Team A's game plan by shutting down its running game.

This is exactly what happens to organizations trying to execute their business strategies and operational plans. To complicate matters even more, like football teams, organizations can be their own worst enemies when it comes to executing plans. Football teams hurt themselves by fumbling the ball, missing blocks, throwing pass interceptions, and missing tackles. The corporate equivalent of fumbles, missed blocks, pass interceptions, and missed tackles can undermine the execution of plans just as effectively as the efforts of competitors. Consequently, you will want to learn how to be "sure-handed" and dependable when carrying out a plan or any portion of a plan for which you are responsible.

Another factor that can undermine the effective execution of plans is what I call the *dirty hands syndrome*. This syndrome is a reluctance on the part of people to step outside of the safe and comfortable intellectualism of planning, and get their hands dirty sorting through the often-messy details of execution. Such people are like the engineer who is comfortable sitting in an air-conditioned office designing a building, but is decidedly uncomfortable when it is time to put on a hard hat and visit the job site to check on the messy details of construction.

Planning involves considering the possible, envisioning the ideal, and looking to the future. It is an appealing intellectual activity full of promise and hope. Execution, on the other hand, involves digging into the details, dealing with reality, and focusing on the here and now. It is a practical, roll-up-your-sleeves-and-get-your-hands-dirty activity that is tempered by the practical realities of business and life. Small wonder, then, that so many people find planning more appealing than execution; a fact that can give you an enormous boost up the career ladder provided you are willing to roll up your sleeves and get your hands dirty.

John and Pete are both managers in a large engineering firm. They have similar backgrounds, education, and experience, but are experiencing vastly different career trajectories. John's career is stalled at the departmental level while Pete has become vice president at a relatively young age. The major difference between these two engineering professionals is *execution*. Ironically, of the two, John is the better planner. He is an outstanding planner while Pete is just a good planner—not great,

SUCCESS TIP

Before beginning the actual execution of a plan, whether strategic or operational, it is a good idea to conduct what I call a *roadblock analysis*. The roadblock analysis consists of an assessment of what might go wrong that could inhibit the effective execution of the plan.

but good. Consequently, it is ironic that Pete's career has had a rapid and steady upward trajectory while John's has leveled off at the department level.

John can think of every base that needs to be covered when developing a plan, but when it comes to executing the plan his attitude could be summarized as follows: "That's somebody else's job." Pete, on the other hand, will be thinking of assignments, deadlines, checkpoints, and roadblocks even as he develops his plans. Because Pete's thinking goes beyond the theoretical aspects of planning to the everyday realities of the workplace, his plans always seem to be executed more effectively than John's. As a result, Pete has left John in his wake as he has climbed up the career ladder.

STRATEGIES FOR EFFECTIVELY EXECUTING A PLAN

Before beginning the actual execution of a plan, whether strategic or operational, it is a good idea to conduct what I call a *roadblock analysis*. The roadblock analysis consists of an assessment of what might go wrong that could inhibit the effective execution of the plan. This assessment consists of answering the following questions:

- What factors might inhibit our efforts to fully implement this plan correctly and on time?
- Do we have the resources to fully implement this plan correctly and on time?
- Do we have the expertise to fully implement this plan correctly and on time?
- Are there existing habits, attitudes, procedures, or informal ways of doing things that might inhibit the effective implementation of this plan?

Once the roadblock analysis has been completed, the following strategies can be used for overcoming obstacles identified and for implementing the plan.

Develop Action/Assignment Sheets

Plans contain goals that are written in broad terms. Your organization will have certain resources available for pursuing these goals. For example, the company has people, processes, and technologies. But even with these resources, there is still an essential ingredient missing: a device for transforming the goals into specific activities, projects, and tasks that can be assigned to responsible personnel and placed within a timeframe.

SUCCESS TIP

Executing a plan is like rolling a ball uphill. It's tough going at first, but if you can keep pushing until you reach the crest of the hill, gravity will take over and it's all downhill from that point on.

The concept of the *transformational device* is best illustrated by an analogy. Assume you are the CEO of a successful engineering company that has an opportunity to grow. Consequently, you adopt the following goal:

Construct a new corporate facility on a fifty-acre tract of land outside of town.

You have identified a contractor who can do the work. This contractor has all the resources needed to construct the facility (e.g., raw materials, people, technologies, processes). However, there is one critical ingredient missing—an ingredient without which the contractor cannot construct the facility. The missing ingredient is a comprehensive set of construction drawings.

The contractor needs the goal to be transformed into operational terms. The device for doing this is a comprehensive set of plans containing the architectural and engineering drawings. The plans give the contractor the specific, detailed information he needs to construct the facility and, thereby, achieve the goal.

Like the contractor in this analogy, people who will be assigned responsibility for executing parts of a plan need a set of "engineering drawings" that transforms their responsibilities into operational terms. These are not real drawings, of course, but they do have all the characteristics of architectural and engineering drawings in that they are detailed and specific. I call these *drawings* for executing plans *action/assignment sheets*.

Action/Assignment Sheets

Action/assignment sheets contain all of the specific activities that must be completed in order to achieve a given goal, the person or group to whom the activities are assigned, and projected completion dates for all activities. Action/assignment sheets show all concerned what tasks must be done, who is responsible for completing each task, and when each task must be completed.

Maintain Momentum by Reinforcing Progress

Executing a plan is like rolling a ball uphill. It's tough going at first, but if you can keep pushing until you reach the crest of the hill, gravity will take over and it's all downhill from that point on. When implementing plans, the role of *gravity* is played by momentum—the tendency of an organization to keep going in a given direction once given a good start in that direction.

Once a sufficient number of action/assignment sheets have been developed and are being worked on, execution of the plan will begin to pick up momentum. If the organization can maintain this momentum, the "execution ball" will make it over the crest and begin the easier journey downhill. To establish and maintain this level of momentum, apply the following strategies as soon as you become a team leader.

- *Model positive execution behaviors.*   While personnel at various levels in your organization are working on their assignments, it is important that they see executives, managers, supervisors, and team leaders working on theirs. Employees will take their cues from those above them in the organization. If they see executives, managers, supervisors, and team leaders putting in the extra effort and time needed to carry out assignments, they will be more inclined to do so themselves.

- *Talk about execution.*   Those leaders with responsibility for an action/assignment sheet should talk up execution. Any time employees who have assignments are gathered together, the responsible leader should talk about the progress being made in completing assigned tasks. It should be obvious from the comments of leaders that executing the plan is a high priority. By talking about the execution process, leaders keep stakeholders focused on completing their assignments.

- *Monitor progress.*   Action/assignment sheets make it easy for those responsible for them to monitor the progress of execution on a daily basis. It is important to do this. This strategy can be combined with the previous strategy—talk about execution. Those personnel with specific assignments should keep copies of their action/assignment sheets close at hand. They should talk to every person with related tasks on their action/assignment sheets every day and not just to ask about progress, but also to see actual evidence of progress. Even the most thoroughly thought-out assignments will overlook something. Even the most determined personnel will run into problems. By monitoring progress daily, responsible leaders can help those with assignments get around the execution problems that inevitably come up so that work doesn't have to stop while the person with the assignment in question wonders what to do.

- *Reinforce results by celebrating.*   One of the best ways to maintain momentum is to reinforce progress by celebrating it. This can be done in a number of different ways. The quickest and one of the most effective is for leaders at all levels to give public "attaboys" to personnel who complete their assignments. Letters of congratulations from senior managers, reports in company newsletters, small-group meetings in which personnel are recognized for completing assignments, periodic written progress reports circulated among all employees recognizing the good work of personnel and teams that complete assignments, and companywide luncheons in which people and progress are recognized are just a few of the many ways progress can be reinforced through celebration.

You can use these strategies to help your team, unit, department, division, or company execute its plans. The better you become at facilitating the effective execution of plans, the further you will go in your career.

REVIEW QUESTIONS

1. Use your own words to define the term *effective execution*.
2. Why do good plans so often fail?
3. Explain the concept of the *roadblock analysis* and the role it plays in the execution of plans.
4. Explain the concept of the *action/assignment sheet* and the role it plays in the execution of plans.
5. Explain how you can maintain momentum once an effective execution is underway.

DISCUSSION QUESTIONS

1. Susan Sims was the best planner at ABC Engineering, Inc. Her department always had the best operational plan every year, but the department's actual performance was never as good as the plan. Discuss some reasons why the plan might be better than the actual performance.
2. Your department has just completed the development of an operational plan. Now you want to know what factors or unforeseen circumstances might inhibit the implementation of the plan. Discuss how you would go about conducting a roadblock analysis for your plan.
3. Your boss wants to implement a new departmental plan, but does not want to bother with action/assignment sheets. Discuss some of the problems the department might confront in trying to execute a plan without action/assignment sheets.

CHAPTER FIFTEEN

Become an Effective Public Speaker

My father gave me these words on speech-making: Be sincere . . . be brief . . . be seated.[1]
   James Roosevelt

The successful professionals interviewed during the study that led to the development of this book were unanimous in their conviction that public speaking is an essential skill for success in any technical career, including engineering. The ability to make presentations to potential customers, represent your company at civic functions, and speak at professional conferences is essential to those who want to advance to the top in engineering.

What made the feedback of these successful professionals even more interesting is that they were almost unanimous in revealing that public speaking did not come easily to them. Rather, it is a skill they had to work hard to develop. Further, the majority of those interviewed said that not only did they dislike public speaking at the beginning of their careers, but they were scared about it. Getting up in front of an audience was a frightening experience for them. Consequently, in order to become accomplished public speakers, the study subjects not only had to develop a new skill, but had to overcome fear.

The feedback from study subjects rang a familiar bell. As a college student, I dreaded the very thought of getting up in front of an audience and speaking. But like you, I had to make class presentations. My choice was simple: Either get comfortable with public speaking, or forget about being a college graduate. It was only with concerted effort and constant practice on my part coupled with commendable patience on the part of some excellent college professors that I conquered my fear of public speaking. Since my college days, I've come to love public speaking, and now look forward to every speaking opportunity.

In fact, public speaking now plays a major role in how I make my living. I speak publicly more than one hundred times a year to audiences large and small. Like the successful professionals interviewed prior to writing this book, public speaking has been an essential skill for me. Without the ability to speak comfortably and well in public settings, my career—like those of my study subjects—would never have developed past the intermediate level.

No matter how you feel about public speaking right now—even if you dread the very thought of it—you can learn to do it well. If you are already comfortable with

SUCCESS TIP

> Those who are not accomplished public speakers tend to think that those who are can simply walk to the podium and deliver an extemporaneous address that rolls effortlessly off the tongue with oratorical eloquence. This is a myth. The number of people who can speak well without preparation is infinitesimally small.

public speaking, you can learn to do it even better. This chapter contains numerous strategies that even the most reluctant person can use to become an accomplished public speaker. These strategies apply regardless of the type of public speaking in question: speech, presentation, demonstration, or testimonial. These strategies are arranged in two broad categories: *preparation* and *presentation*.

PREPARATION: THE ESSENTIALS

Those who are not accomplished public speakers tend to think that those who are can simply walk to the podium and deliver an extemporaneous address that rolls effortlessly off the tongue with oratorical eloquence. This is a myth. The number of people who can speak well without preparation is infinitesimally small. In fact, some of the best-known orators of modern times—Winston Churchill, Martin Luther King, Jr., John F. Kennedy, and Ronald Reagan—worked hard to prepare their speeches. Churchill actually wrote out every word he planned to say and even noted in his speech where to pause for emphasis. For most people, it takes a lot of preparation to give a speech that appears to be extemporaneous.

Effective public speaking for most people is seventy-five percent preparation and twenty-five persentation. The preparation steps are as follows: (1) prepare yourself, (2) prepare your materials, (3) select the appropriate technology (if any), (4) prepare the facility, and (5) prepare your audience. Each of these preparation steps is covered in detail in the following sections.

Preparing Yourself

Preparation is the key to effective public speaking. Never speak in public without first preparing yourself. Public speaking is like any other human endeavor in that the better

SUCCESS TIP

> Preparation is the key to effective public speaking. Never speak in public without first preparing yourself. Public speaking is like any other human endeavor in that the better you prepare, the better the result.

you prepare, the better the result. If you are a fan of sports, you know that every player's preparation begins well before the game or match. Even on the day of the game, baseball and softball players arrive at the stadium hours before game time to take batting practice, stretch their muscles, take fielding practice, have ankles and wrists wrapped by the team's trainer, and review scouting reports about the other team. Men and women who play other sports follow a similar routine. It's all about preparation.

When asked to make a speech or presentation, follow the lead of athletes—prepare yourself. What follows are some strategies that will prepare you to speak with confidence:

- *Do your research.*   The first step in preparing to speak involves conducting any and all necessary research into your topic. If you have been asked to speak, you are probably already well versed concerning the topic in question. However, there are likely to be people in the audience who, like you, know a great deal about the topic in question. Consequently, no matter how much of an expert you are concerning the topic of your address, review the latest literature and other applicable material to make sure you are up-to-date. This can save you the embarrassment of being asked a question you should be able to answer but can't or, worse yet, giving an incorrect answer.

- *Say "no" when appropriate.*   I may be jumping the gun a little in giving this advice so early in this chapter, but the advice is critical. Once you become an accomplished public speaker, you will begin to receive invitations to speak from a variety of different sources (e.g., civic clubs, professional organizations, colleges, schools). Groups such as these are constantly seeking good speakers, and they depend on local companies and organizations to provide them. They typically have program chairs or other people responsible for identifying and securing speakers. If one of these program chairs has heard you speak and was impressed, you are likely to get a call. When this happens, if you are not well versed in the topic that you are asked to speak about, just say "no," or at least suggest an alternate topic. Even the most articulate, refined, and eloquent orators cannot give credible speeches on topics about which they know little or nothing.

- *Plan to use humor appropriately, but don't be a stand-up comic.*   Humor can add spice to a speech or presentation and improve its reception by the audience. However, don't forget that you were asked to speak based on your expertise in a given field. Use humor to enhance your speech in the same way you use salt and pepper to enhance the taste of a meal—in just the right amounts. Laughter and approval from an audience are powerful, alluring forms of positive feedback. They can have the same effect on you as drugs—once you've had a little, you'll probably want more and more and more. Many people become so addicted to the laughter they receive from audiences that humor begins to replace substance and content in their speeches. This is a serious mistake—one that won't help your career unless, of course, you plan to be a stand-up comic. When preparing your speech or presentation, build in humor sparingly to emphasize points and to get the audience's attention, but not to replace substance and content.

■ *Select a comfortable outfit.*   When you first start making speeches or presentations, you will probably be nervous if not downright mortified. This is normal. If you are like most people, public speaking will always make you a little uncomfortable—even after you get good at it. Lots of people still get "butterflies" in their stomach before speaking in public, even after they are accomplished orators. If this is the case with you, don't do anything to add to your discomfort. For example, be sure to wear a comfortable outfit when making a speech or presentation, one that is appropriate for the occasion and season, but that is also comfortable to wear. You don't want to be self-conscious about what you are wearing or uncomfortable because your clothes are too tight, too loose, or too anything else that might add to the discomfort you already feel about speaking in front of an audience.

■ *Eat less than usual before speaking.*   Many speeches and presentations are preceded by a meal—usually lunch or supper, but breakfast meetings with a speech are becoming more and more common. If you are to be the speaker in one of these situations and you are at all nervous about speaking, eat less than you normally would. I was present at a meeting when the after-dinner speaker was so nervous that before he could finish the first line of his speech, the unthinkable happened—he got sick and deposited his meal in the laps of his listeners. This is an extreme example, but the point is still valid. If you get nervous before speaking, do not fill your stomach with food. Instead, cut back. Eat about half of what you normally would.

■ *Have a glass of water handy.*   When you walk to the podium, either take a glass of water with you or arrange to have one already placed there. If you get a case of dry-throat or cottonmouth while speaking, a sip of water can be a lifesaver. Notice that I said water—not tea, soda, coffee, or any other caffeinated drink. Caffeine is a diuretic. Drinking caffeinated drinks just before or during a speech can cause a most uncomfortable and embarrassing situation to arise in the middle of your speech.

■ *Visit the restroom.*   Arrange to discreetly visit the nearest restroom just prior to speaking. This is an extension of the previous strategy. Even if you don't feel the need, visit the restroom before beginning a speech or presentation. I have been present in the audience when a speaker had to ask participants to take a break so he could ". . . answer a cell phone call." It was noted with amusement by many in the audience that he took the call in the restroom. Needless to say this speaker's credibility suffered in the eyes of his listeners.

■ *Write your own introduction.*   When someone asks you to make a speech, they will typically ask for a copy of your resume so that someone can introduce you. Don't ever give someone your resume to use in making an introduction. Instead, write out a short introductory biography and have it typed in a large font with boldface lettering. To the extent possible, tailor your biography to the audience in question. Give it to the person who is to introduce you—with your name spelled out phonetically if it is a difficult name—and ask that person to read it verbatim when introducing you. I cannot remember how many times I've seen a speech get off to a bad start—but it's many—because the person who is supposed to introduce the speaker doesn't even look at his resume until actually walking to the

SUCCESS TIP

> Until you are sufficiently accomplished to have a sense of the time when speaking, it is best to plan what you are going to say and time how long it takes you to say it. I do NOT recommend, however, that you take the entire written speech to the podium with you. Rather, I recommend that you take a comprehensive outline containing numbered or bulleted items followed by key terms and phrases that will jog your memory concerning what you want to say. If you take a written-out speech to the podium, you will probably wind up just reading it, and no audience wants to be read to.

podium. Then he begins frantically riffling through the unfamiliar document looking for all the world like the unprepared host he really is. Then, just to make sure things get off to a really bad start, the unprepared host mispronounces the speaker's name. When this embarrassing situation had happened to me twice, I decided to take matters into my own hands. I don't just write out my own introductory biography complete with the phonetic spelling of my name, I send it to the host ahead of time and bring another copy with me. Then, before the host goes to the podium to introduce me, I have him show me the biography I provided and I remind him how to pronounce my name. If he suddenly has a confused expression and starts rummaging through his coat pockets looking for the biography, I give him the extra copy I brought and tell him to use it. This might seem to be overly assertive to you, but take my word for it, with public speaking, it's better to be assertive and sure than timid and sorry.

■ *Develop your outline or notes.* When you are first beginning to make public speeches or presentations, you might need to write down everything you plan to say. This will serve two purposes. First, it will force you to give an appropriate level of forethought to what you plan to say. Second, it will allow you to time your speech. Most speeches and presentations have a definite time limit (e.g., thirty, forty-five, or sixty minutes). Until you are sufficiently accomplished to have a sense of the time when speaking, it is best to plan what you are going to say and time how long it takes you to say it. I do NOT recommend, however, that you take the entire written speech to the podium with you. Rather, I recommend that you take a comprehensive outline containing numbered or bulleted items followed by key terms and phrases that will jog your memory concerning what you want to say. If you take a written-out speech to the podium, you will probably wind up just reading it, and no audience wants to be read to.

■ *Practice your speech.* Until you become accomplished enough to do otherwise, read your written-out speech numerous times until you can remember what you want to say without even thinking about it. Then, develop your annotated outline and practice giving the speech or making the presentation using just the outline. Practice in this manner until what you plan to say comes naturally and is second nature. Then and only then are you ready to give the speech or make the presentation in public.

Success Tip

One of the most enduring rules of public speaking is this: *Whenever possible put something in their hands.* This means that whenever possible, you should give every person in the audience a hard-copy outline or summary of what you are going to say.

Preparing Your Materials

When making a speech or presentation, you don't want to be just a talking head. Your presentation will be received more favorably if you make it visual. This means that unless it is impractical to do so, you should provide members of the audience some type of handout (e.g., an outline, supportive graphics, a copy of the speech). This is true even when you make an audiovisual presentation (e.g., PowerPoint, slides). In such cases, give the audience copies of your visuals.

When you see the president of the United States make a speech on television, all you see is a person talking while standing at a podium or sitting at a desk. What you don't see is that all members of the press corps have already been given copies of the president's speech, or at least a summary of it. This is important. One of the most enduring rules of public speaking is this: *Whenever possible put something in their hands.* This means that whenever possible, you should give every person in the audience a hard-copy outline or summary of what you are going to say. Regardless of what you give listeners, it is important to provide something—preferably something that will help listeners follow along as you speak, jot down notes if they wish to, and remember what you said when the speech is over.

If you are going to provide handouts to the audience—and you should whenever possible—it is important that the handouts be well prepared, attractive, and a positive reflection of the image you hope to project. Keep the following strategies in mind when preparing materials to support a speech or presentation:

■ *Keep materials simple.*   Overly complicated visuals that are difficult to read are worse than no visuals at all. The purpose of visual aids is to simplify and summarize otherwise complex information. If you plan to use tables or charts, keep them simple and easily readable.

■ *Use bullets and key statements.*   I never give the audience a copy of the full text of a speech or presentation. Do this and some members of the audience will skip your presentation and just read it later. Instead, I recommend distributing an annotated outline, perhaps even the one you developed for yourself. Bulleted items with key statements make a brief but comprehensive handout that can serve as both an outline and a place for taking notes.

■ *Think big—prepare visuals as if audience members have poor vision.*   Visuals serve no purpose if the audience can't see them. In fact, providing visuals that cannot be easily read is worse than providing no visuals at all. For visuals that will be distributed in hard copy or presented on a screen, choose a large type size and a bold font. Make sure that a person who doesn't see well can see your materials.

SUCCESS TIP

My best advice for people preparing to make a speech or presentation that is supported by technology is this: *Don't let the tail wag the dog.* You want listeners to remember what you say, not the technology you used to say it.

■ *Build in redundancy.*   Whenever asked to make a presentation, I develop a Power-Point disk, overhead transparencies, and hard copies. In this way, no matter what goes wrong—and things often go wrong—I can still make my presentation. For example, if a computer glitch prevents the presentation file from opening, I will switch to the overhead projector. If the projector breaks down, I will work from the hard copies. As soon as you begin to make presentations and speeches, you will understand the value of redundancy in the preparation of support materials.

■ *Make more copies than you expect to need.*   If your host tells you to expect fifty people in the audience, bring sixty copies of your handouts. The number of times I've seen speakers run out of handouts before everyone in the audience has one is so high I've stopped counting. When this happens, those audience members who don't get a handout immediately become disgruntled, and even those who do get copies sympathize with those who don't. This is a bad way to begin a speech or presentation. It makes you appear to be unprepared and turns the audience against you from the outset.

Select the Appropriate Technology

If you are accustomed to computer animation and other similar presentation technologies, what I say in this section might surprise you. My best advice for people preparing to make a speech or presentation that is supported by technology is this: *Don't let the tail wag the dog.* You want listeners to remember what you say, not the technology you used to say it. To understand what I mean, think of television commercials. Some are so well produced and entertaining that you forget the product being advertised and just remember the commercial. This is not the result desired by the company that paid for the commercial.

Don't let a similar fate befall your speech or presentation. I have seen many speakers use Hollywood-quality graphics and animation during presentations only to have the audience become so wrapped up in the technology that they miss the speaker's message. Once, after watching one of these hyper-tech presentations, I heard the following conversation between two audience members: "The speaker really had great graphics, didn't he?" "He sure did. I don't know what the presentation was about, but the graphics were outstanding." Unless this speaker is marketing computer graphics technologies—which he wasn't—his presentation would fail, and it failed, miserably.

Other cautions to observe when choosing the technology to support your speech or presentation are as follows:

■ *Make sure you can operate the technology.*   I've seen many presentations get off to a bad start because the speaker didn't know how to operate the technology he planned to use. If you cannot operate the technology you plan to use without technical assistance from a third party, make absolutely sure the third party will be available. If there is any doubt about the availability of technical assistance, either learn to operate the technology yourself or choose another technology you can operate.

■ *Make sure the facility will accommodate the technology you plan to use.* If you plan to use devices such as laptops and LCDs when making your speech or presentation, make sure the facility in question will accommodate the devices. For example, is there a screen available? Are there electrical outlets nearby? Can the lights be dimmed sufficiently to allow for the projection of a clear image? Will the seating arrangement allow audience members to easily see the images you plan to project? If the answers to these questions are negative, can the facility be rearranged to accommodate your technology? If not, use another approach.

■ *Make sure your speech or presentation can still be given if the technology fails.* Over the years, I've seen a lot of speakers embarrassed and audiences disappointed because a speech or presentation had to be cancelled when the speaker's technology broke down, malfunctioned, was lost by the airline, or got left on the seat of a taxicab. When you plan to use technology to support a speech or presentation, make sure you have a way to continue with the presentation in spite of technological glitches. Remember my earlier advice about redundancy. If you have a good outline and a set of handouts, you will always be able to make your speech or presentation regardless of technological glitches.

Preparing the Facility

When considering the facility where you will make your speech or presentation, remember two things: arrangement and accommodations. By arrangement I mean how the room is set up. How is seating for the audience arranged? Will everyone in attendance be able to hear you and will they be able to see your visuals? By accommodations I mean lighting, electrical outlets, podium, screen, microphone, and anything else you will need in order to make the presentation go according to your plan.

A conducive arrangement and accommodations can enhance the quality of your presentation. However, an arrangement that is not conducive to presentations can ruin yours, no matter how well you deliver it. Consequently, I recommend the following strategies relating to the facility:

■ *Conduct a "reconnaissance" mission ahead of time.* If possible, visit the facility in advance of your presentation. If this is not possible, contact the person hosting your talk and discuss facility issues. I typically e-mail instructions to the host expressing my wishes concerning how the facility should be set up. Then, on the day of your presentation, arrive early enough to make changes if the facility's arrangement is not conducive to what you plan to do. For example, if the seating is arranged in theater style and you want the audience seated in small groups, arriving early will give you time to have the necessary changes made or make them yourself. There have been many times when I simply made the changes myself. If you don't like the location of the podium, screen, projector, or any other support equipment you plan to use, have it moved or move it yourself.

- *Determine how to operate the lighting.*   Can the lights in the room be dimmed? If so, where are the dimmer switches and can you operate them yourself or ask someone in the audience to do it? Are there blinds or drapes over any windows that might need to be closed? How are they operated? Try making lighting adjustments before the audience arrives.

- *Locate the microphone.*   Will you need a microphone in order for everyone in the audience to be able to easily hear everything you say? If so, is one available? Is it attached to the podium or is it a cordless portable model? How do you turn the microphone on? How do you adjust the volume on the microphone? Make sure the microphone works the way you need it to and that you know how to work it without help from a technician (who might not be available when needed).

- *Locate the podium.*   Do you plan to use a podium? If so, is there one available? Is it located where you need it to be? Can it be moved? Does it have a reading light on it that can be turned on when the lights in the room are dimmed? How is the light operated?

- *Locate the electrical outlets nearby.*   Will you need to plug in a laptop computer, projector, or any other equipment in support of your presentation? If so, are there sufficient electrical outlets nearby? Will you need an extension cord? Is there a cord available? Are the plugs on all electrical cords compatible with the electrical outlets?

- *Locate flip charts and marker boards.*   Will you need to use a flip chart or marker board during your presentation? If so, is there one available? Will everyone in the audience be able to see it? Are there markers available? How about an eraser for the marker board? My advice is that if you will need markers, bring a supply with you. The ones provided might be dried out or the wrong kind.

- *Determine if the temperature in the room can be controlled.*   One of the most persistent facility problems for speakers is room temperature. Invariably, rooms are either too hot or too cold. In either case, temperature problems can make it difficult to hold the audience's attention. Most facilities have the temperature set for comfort when the room is empty. When an audience packs in, normal body heat can raise the temperature in the room noticeably. Ask the host to have the facility manager set the temperature slightly cooler than usual to compensate for the body heat of the audience. What's even better is to find the thermostat and determine how to adjust the temperature yourself.

- *Determine if there are distractions that can be eliminated.*   One of the benefits of arriving early is that you gain the time necessary to identify and eliminate distractions. For example, I once made a presentation at a well-known resort in central Florida during the summer months. The facility was excellent except for one thing. It had floor-to-ceiling windows looking out onto a pool that was packed with sunbathers. Fortunately, the windows had drapes and I was able to find someone who knew how to close them. The presentation was well received, but I don't think it could have competed with the swimming pool distraction had the windows not been draped. If your presentation is going to be an after-lunch or after-dinner talk, arriving early will give you time to locate the head waiter and ask that serving staff hold off picking up dishes until after you finish speaking or, at the very least, that they pick the dishes up quietly. I have given after-lunch and after-dinner talks during which the noise from waiters clanking dishes and silverware was so loud the audience couldn't hear me even with the volume control turned up. Another frequently-encountered distraction is noise from any number of different sources (e.g., renovations being done to the building you are in, construction on nearby sites, a meeting taking place in the room next door, squeaky heating or cooling equipment). There will be times when nothing can be done to eliminate distractions. However, it has been my experience that most of the time you can eliminate distractions if you arrive early enough to identify them. I've asked facility personnel to move my presentation to another room to get away from noise distractions. Arriving early at least gives you the time needed to eliminate distractions.

Preparing the Audience

Preparing the audience means getting them interested in what you have to say, eager to hear you say it, and ready to listen. There are several strategies that can be effective for preparing listeners. If the audience members receive flyers or registration material in advance of your talk, help your host make the announcement about your presentation interesting and attractive. Ask to see the draft of the flyer containing the announcement for your presentation before it is sent to the printer. Also, provide biographical information for inclusion in flyers and registration material that is tailored to appeal to the audience in question. You want your biographical information and the announcement about your presentation to gain the attention of potential listeners so powerfully they will say, "I want to hear this person." Once the audience is in the room, seated, and ready to hear what you have to say, spend a few moments doing what entertainers call "warming up the audience." Don't just plunge right into your talk. Instead, open with the public-speaking version of small talk. Most books and organizations that teach public speaking advise you to use a joke to warm up the audience. I differ strongly with this advice for two reasons. First, not everyone can tell jokes well. If you are one of those who can't, your attempt will just make you and the audience uncomfortable. Second, even if you are an accomplished joke teller, in a diverse society such as ours it is difficult to find a joke that won't offend someone. A joke that offends people in the audience will alienate the very people you need to win over. Rather than tell jokes, I recommend that speakers tell

SUCCESS TIP

Most books and organizations that teach public speaking advise you to use a joke to warm up the audience. I differ strongly with this advice for two reasons. First, not everyone can tell jokes well. If you are one of those who can't, your attempt will just make you and the audience uncomfortable. Second, even if you are an accomplished joke teller, in a diverse society such as ours it is difficult to find a joke that won't offend someone. A joke that offends people in the audience will alienate the very people you need to win over.

Success Tip

Be enthusiastic about your topic. Why should the audience be interested in what you have to say if you aren't? If you show an interest in your topic, the audience will usually respond in kind.

an amusing or heart-warming story—preferably one that relates to the subject of your presentation. The public-speaking section of any major bookstore will contain numerous books replete with stories for speakers. Once you begin receiving invitations to speak publicly, it is a good idea to add one or two of these books to your personal library. Finally, once you have warmed up the audience, review the importance of what you have to say—but do this from their perspective. Ask yourself the following question when preparing your presentation: *Why would someone come to hear me speak on this subject?* Use the answer to this question to develop a brief statement that will gain the attention and interest of your audience.

PRESENTATION: THE ESSENTIALS

There are three presentation methods I recommend for public speaking. I call these the stand-and-deliver, storytelling, and Socratic methods. You can use one or any combination of all of these. Regardless of which method or combination of methods you choose, there are numerous presentation strategies that apply. I call these my "universal presentation strategies" because they apply regardless of the presentation method you choose.

Universal Presentation Strategies

Regardless of the type of presentation you plan to make and regardless of the audience that will receive the presentation, the following strategies apply

- *Maintain eye contact with the audience.*   Some books and organizations that teach public speaking recommend looking just over the heads of the people you are speaking to. This is supposed to help relieve some of the anxiety you feel about being in front of an audience. I disagree with this advice. Eye contact with the entire audience is critical. It's how you connect with your listeners, and connecting with listeners is key to a successful presentation or speech. Just think about how you feel when talking with people who will not look you in the eye. You probably don't trust what they are saying. Rather than adopt some artificial trick to relieve anxiety, use the strategies set forth in this chapter to get comfortable speaking to an audience. Look your audience members in the eyes and spread your eye contact around continually to include the entire audience.
- *Be enthusiastic about your topic.*   Nothing will turn an audience off more quickly than a presentation delivered in a lifeless monotone. People who speak in a monotone

devoid of interest and enthusiasm risk losing their audience or at best boring it into tedium. Be enthusiastic about your topic. Why should the audience be interested in what you have to say if you aren't? If you show interest in your topic, the audience will usually respond in kind.

■ *Be yourself.* I know a speech teacher who once insisted that every student place both hands flat on the podium and never move them during their entire presentation. As a consequence, her students looked like granite statues when making a speech. Within some common-sense limits, it's better to find your own style—a style of speaking you are comfortable with and a style that is *you*—than to adhere to someone else's one-size-fits-all guidelines. If you "speak with your hands" in normal conversation, feel free to do so when making a speech or presentation. Don't go too far and look like a wounded bird flapping its wings, but do use your hands to emphasize selected points if that is the style you are comfortable with. If you prefer to move around rather than stand behind a podium, do so as long as you have the room and a portable microphone. Just remember, as you move around, maintain eye contact with your audience. I once saw a speaker who paced frantically back and forth while speaking and never even looked at the audience. His style violated the common-sense limits I mentioned earlier. However, some movement while maintaining eye contact with the audience can keep you from becoming just a talking head behind a podium.

■ *Don't be a pompous windbag.* No matter how well you know your subject and no matter how accomplished you become at public speaking, never fall into the trap of becoming a pompous windbag. Let people ask questions and take their questions seriously—even if they seem elementary to you. You are speaking to educate and inform, not to pontificate and dominate. Be humble. Your audience will like you better that way.

■ *Don't say you know when you don't.* It's going to happen. You are going to be asked a question you cannot answer. When this happens, it is best to try one of the following strategies: (1) Say, "Good question. Is there anyone in the audience who can answer it?" or (2) Say, "I don't know, but if you will leave me your business card after my talk I'll find out for you." If you give an inaccurate answer, the credibility you've worked so hard to establish will be lost. You don't have to know everything. The audience will appreciate your desire to be accurate when answering questions.

■ *Allow questions and listen to them carefully.* Unless the format simply does not allow it, let audience members ask questions. Questions can be taken during or at the end of your presentation. You should decide ahead of time how you will handle

SUCCESS TIP

Don't get into a shoving match with self-promoting know-it-alls in the audience. Either ignore them or acknowledge their comments with the word *interesting* and move on to someone else. The audience will know what you mean by *interesting*.

Success Tip

There is an old saying about public speaking that goes like this: *Tell 'em what you're going to tell 'em, then tell 'em, then tell 'em what you told 'em.* This is just a folksy and easy-to-remember way of saying have an opening, body, and conclusion.

questions and let the audience know during the opening portion of your talk. When someone in the audience asks a question, listen carefully. Then before answering, repeat the question to ensure that other audience members heard it and that you heard it correctly. Never allow a listener's question to make you angry. When speaking in front of an audience, it is even more critical than ever to maintain your emotional equilibrium. Speakers who let questions from the audience cause them to lose their temper will also lose their credibility. Self-promoting hecklers should be either ignored altogether or given only the most cursory response such as "That's an interesting comment. Does anyone else have a question or comment?" Don't get into a shoving match with self-promoting know-it-alls in the audience. Either ignore them or acknowledge their comments with the word *interesting* and move on to someone else. The audience will know what you mean by *interesting.*

■ *Eliminate all distracting affectations and habits.*   As you develop your public-speaking skills, ask a trusted friend or colleague to observe and make note of any distracting affectations or nervous habits you exhibit while speaking (e.g., jiggling coins or keys in your pockets, drumming your fingers on the podium, running your fingers through your hair, twisting a lock of hair, rocking back and forth). Also ask your friend to make note of speech affectations such as "uh," "ah," "you know," "like," etc. If possible, have your friend videotape your early presentation. You will be surprised—*shocked* and *embarrassed* are probably more appropriate words here—to see some of the things you do. Just make a mental note of these distracting behaviors and work on eliminating them the next time you speak.

■ *Don't rush.*   If you are nervous about public speaking, your tendency will be to rush. If this is the case with you, make a conscious effort to slow down. People in the audience will have enough trouble following what you say and remembering it. Don't add to their difficulty by speaking too fast. A relaxed pace is better for listeners.

STAND-AND-DELIVER METHOD

This is the method being used when someone makes a speech. In college classes, you probably thought of it as the lecture method. Although the stand-and-deliver method is often a one-way broadcast of information, it doesn't have to be. In fact, stand-and-deliver speakers who encourage listeners to ask questions or offer comments can improve the response they get from the audience, even if questions and comments are held off until the end of the talk.

Stand-and-deliver presentations and speeches should have three distinct elements: (1) opening, (2) body, and (3) conclusion. A fourth element—questions/comments—is optional, but recommended. There is an old saying about public speaking that goes like this: *Tell 'em what you're going to tell 'em, then tell 'em, then tell 'em what you told 'em.* This is just a folksy and easy to remember way of saying have an opening, body, and conclusion.

Opening

This is the part of the speech or presentation in which you warm up the audience with a humorous or otherwise interesting story. Having done that, you then give listeners a brief overview of what you plan to talk about. This brief overview need be nothing more than a statement of the major topics your speech or presentation will cover. Providing an overview lets listeners know what to expect. It also provides an opportunity to get the audience interested in what you are going to say.

Body

This is the part of the speech or presentation in which you actually present the material you planned to share with the audience. Remember to stick to your outline and concentrate on pacing yourself so listeners can keep up. Stories that illustrate selected points will help the audience remember, and they will help you hold the attention of your audience.

Conclusion

This is the summary portion of your speech or presentation. At this point you restate the title and purpose of the talk. Then you briefly summarize the key points you have made. If you have drawn any conclusions or recommended any action, reiterate them here. Once you have completed the conclusion portion of the speech or presentation, open the floor for comments and questions. If there is anything the audience members are supposed to do in the follow-up to your speech or presentation, close with a reminder after all comments and questions have been dealt with.

STORYTELLING METHOD

It is rare that an entire presentation can be conveyed by a story. However, storytelling mixed with the stand-and-deliver method can result in an especially effective speech or presentation. Stories are best used to illustrate selected points. For example, in the speech I sometimes give at college graduations, I make the point—as I do in the last chapter of this book—that perseverance is one of the keys to success. I want the listeners to know that they are going to have some false starts, make some mistakes, and even experience failure, but if they keep trying and refuse to quit they will eventually win—at least most of the time.

To illustrate the value of perseverance, I tell the following story: *There is a man all of you know as one of the most successful, most famous Americans in our nation's history. But some of you might not know how often he failed and had to keep trying in order to become the person we now all know. When this man was just 22, he failed in business. At the age of 23, he ran for a seat in the legislature of his state, but lost the race to another candidate. At the age of 24, he tried again to make it in business, but again failed. At the age of 27, he had a nervous breakdown. At the age of 34 he decided to try national politics and ran for a seat in the U.S. Congress, but lost the race. At the age of 46, he decided to run for a seat in the U.S. Senate, but again he lost the race. At the age of 47, he campaigned to be vice president of the U.S. and was defeated. At the age of 49 he decided to make another run for a seat in the U.S. Senate, but once again he lost the race. A lesser man would have given up long ago and just quit, but not this man. At the age of 51, he decided to run for the country's highest political office—president of the United States. Guess what? This time he won!*

After telling this story, I ask the audience to tell me whose story I've just told. There is usually someone who knows that this is the story of Abraham Lincoln. I could simply tell the graduates that perseverance is important to success, but such a statement would be weak compared with the power of Lincoln's story. This is just one example of how storytelling can be used to enhance a speech or presentation. (Many stories are available that will illustrate almost any point.) For you, it's just a matter of finding, cataloging, and using them.

SOCRATIC METHOD

This was the method Socrates is said to have used in teaching his student Plato. In more contemporary times, it was the method used by quality guru J. Edwards Demming to help the Japanese rebuild their industrial base and emerge from the ashes of World War II as an economic superpower. The Socratic method consists of posing questions and expecting the listener to provide the answers. Rather than relate information as is done in the stand-and-deliver and storytelling methods, you simply ask a series of questions and let listeners come to their own conclusions concerning the answers. This is the important part of the Socratic method. You don't provide answers—only questions. If listeners suggest an answer that is poorly thought out or clearly off the mark, you guide them to that conclusion using other questions.

The Socratic method can be an excellent way to facilitate brainstorming and to force people to think critically. I sometimes use this method when working with management personnel in organizations that are not performing as well as they should. I begin by asking them a simple question: *In a perfect world, what would you like this company's performance to be?* When they have answered this question, I ask another: *Why isn't this company performing the way you want it to?* I record the answers provided on a flip chart or marker board, and continue to ask questions that eventually lead participants to an understanding of what they will need to do in order to improve their company's performance. What is important here is that by using the Socratic method, I create a situation in which the participants solve their own problems rather than just being told what to do by an outside consultant. You can use the Socratic method with your team, unit, department, division, or organization as you move up the ladder throughout your career.

REVIEW QUESTIONS

1. Explain how to prepare yourself for a public-speaking engagement.
2. Explain how to prepare your materials for a public-speaking engagement.
3. When selecting the technologies to support your speech or presentation, what factors should you consider?
4. Explain how to prepare the facility before making a speech or presentation.
5. Explain how to prepare the audience at the beginning of a speech or presentation.
6. List and explain at least five universal presentation tips for making speeches and presentations.
7. What are the three parts of a "stand-and-deliver" speech or presentation?
8. Explain how you can use stories to increase interest in speeches and presentations.
9. Explain the "Socratic method" and appropriate uses of it in public speaking.

 # SUCCESS PROFILE

Teddy A. (Al) Ward
Engineer and Manager

Teddy A. (Al) Ward, an industrial engineer by training, is now the general manager of Florida Transformer, Inc.; a company that rebuilds transformers for electric utility companies throughout the southeastern United States. Ward rose from being a fresh college graduate with an engineering degree to the top operational position in a successful technology company by applying the various strategies explained in this book. Ward is a professional of unimpeachable integrity, a skilled strategic thinker who also knows how to execute, and a creative problem solver. His career especially exemplifies the concepts of teambuilding and the positive use of public speaking.

Ward's goal is to achieve peak performance from his company's people and processes and to improve that performance continually. Consequently, he carefully monitors the performance of the various teams that make up the workforce at Florida Transformer, Inc., making note of performance areas in need of improvement. When he identifies a need for improvement, Ward organizes formal training, mentoring, or some other appropriate form of help to facilitate the needed improvement. For example, Ward became concerned that supervisory personnel in his company typically become supervisors by performing well as engineers, technicians, or in other nonmanagement positions, but lack any formal instruction in supervision (this is true of most technology companies).

Just one example of his efforts will show how Ward uses teambuilding to continually improve the performance of his company. In order to strengthen his supervisors as part of the company's overall management team, Ward purchased copies of the book *Effective Supervision* for all supervisors and hired the author of the book to take them through a comprehensive course on the subject. In order to

(continued)

(*continued*)

provide the training on company time and still maintain his production schedule, Ward had to divide his supervisors into two groups. While one group participated in the training, the other doubled up on the job to ensure that contracted work continued on schedule. Then, the groups changed places.

In this way, the company's supervisors received critical training in such key areas as leading teams, communicating with employees, ethics, motivating team members, problem solving, performance appraisals, handling employee complaints, managing conflict, legal issues (e.g., employee discipline, sexual harassment, drug abuse, and employment termination), and safety on the job. With this management training added to the technical expertise and solid work experience they already had, the company's supervisors are now a stronger component of the company's overall management team.

After completing college, Ward worked in a series of higher and more responsible positions that took him from the United States to Brazil to Korea. His former employers include both Western Electric and AT&T. Once Ward reached the management level, it became necessary for him to represent his employer in professional organizations and civic groups. For example, Ward has served as chairman of the board of a local chamber of commerce, a member of the board of directors of a local Workforce Development Board, and a member of such civic groups as the Jaycees and Rotary International.

As his company's representative to both civic and professional groups, it has been necessary for Ward to develop and hone his public-speaking skills. Although he had to overcome all of the same difficulties most people experience when they first begin to make public speeches and presentations, Ward persevered and became an accomplished public speaker. He has now spoken to business and civic groups throughout the world, and does so with competence and finesse. Ward's example is one that any engineering student would do well to emulate.

DISCUSSION QUESTIONS

1. Assume you have been asked to speak to a local civic club on a topic with which you are not familiar. Discuss ways to handle this situation.
2. Have you ever seen a presentation made that used so much technology that the speaker's message got lost? If so, discuss the situation and explain your reaction to the presentation.
3. Assume your professor gave you the assignment of making a presentation in class. You have prepared yourself, your materials, and all necessary technological support, but the classroom is not arranged well. Discuss how you should handle this situation.
4. Select a topic for a presentation you might make in class. Now discuss how you could use stories to make your presentation more interesting.

ENDNOTE

1. Louis E. Boone, *Quotable Business*, 2nd ed. (New York: Random House, Inc., 1999), 60.

CHAPTER SIXTEEN

Become an Effective Problem Solver and Critical Thinker

It is when solving problems that a leader's ability to innovate is most apparent. It is when making decisions that a leader's accountability is most apparent.

   Anonymous

Effective problem solving and decision making are fundamental to success in your career. On the one hand, good decisions will decrease the number of problems that occur. On the other hand, the workplace will never be completely free of problems. This chapter will help you learn how to solve problems effectively, positively, creatively, and in ways that don't create additional problems; become better at critical thinking; and learn to think, make decisions, and handle problems in ways that will advance your career.

SOLVING AND PREVENTING PROBLEMS

Even the best-led organizations have problems. No matter where you work, there will be problems. A problem is any situation in which what exists doesn't match what is desired, or, put another way, a problem is a discrepancy between the actual and the desired state of affairs. The greater the disparity between the two, the greater the problem. Problem solving is not just "putting out fires" as they occur. Rather, it is a way to make continual improvements in the performance of people and organizations. The better you get at solving and preventing problems, the further you will go in your career.

The Perry-Johnson Method

Perry Johnson, Inc., of Southfield, Michigan, recommends an approach to problem solving that works well because of its three main characteristics:[1]

   1. It promotes teamwork in problem solving.
   2. It leads to continual improvement rather than just "putting out fires."
   3. It approaches problems as normal by-products of change.

Learn to use this method to become an effective problem solver. The Perry-Johnson Method for problem solving is as follows:

■ *Establish a problem-solving team.*   The reason for using a team in solving problems is the same as that for using a team in any undertaking: Teams multiply the abilities of individuals. Team members have their own individual experiences, unique abilities, and particular ways of looking at things. Consequently, the collective efforts of a team are typically more effective than the individual efforts of one person. A problem-solving team can be a subset of one functional unit such as a project team, or it can have members from two or more different units, teams, departments, etc. It can be convened solely for problem solving, or it can have other duties. Decisions about how to configure the team should be based on the need, size, and circumstances of the organization as well as the nature of the problem(s) in question.

■ *Brainstorm the problem list.*   It is important to get out in front of problems and deal with them systematically. For example, the military doesn't just sit back and wait for the next trouble spot in the world to explode. Rather, potential trouble spots are identified and entered onto a problem list. The potential problems are then prioritized and plans are developed for handling them. You can use this approach in your organization. The problem-solving team should brainstorm about problems that might occur and create a master list.

■ *Narrow the problem list.*   The first draft of the problem list should be narrowed down to the entries that are the worst problems. To accomplish this, evaluate each entry on the list by means of three criteria:

    1. Is there a standard to which the entry can be compared?
    2. Does actual performance vary from the standard in an undesirable way?
    3. Is the variance supported by facts?

The answer to all three questions should be "Yes." Any entry that does not meet all three criteria should be dropped from the list.

■ *Create problem definitions.* All problems remaining on the list should be clearly defined. A problem definition has two parts: a description of the circumstance and a description of the variance. For example, here is a problem definition showing the problem, the circumstance, and the variance:

    ■ Problem: Corporate office is angry because Job 21A is over budget by twenty-two percent
    ■ Circumstance: Job 21A is over budget
    ■ Variance: By twenty-two percent

■ *Prioritize and select problems.*   With all problems on the list defined, your team can prioritize them and decide which one to pursue first, second, and so on. Perry Johnson recommends using a Problem Priority Matrix. The matrix is created as follows:

    ■ Divide the problem-solving team into two groups and put them in separate rooms.
    ■ Have Group A rank the problem list in terms of how much a solution will benefit the organization.

- Have Group B rank the problem list in terms of how much effort will be required to solve the problem.
- Set up a Problem Priority Matrix based on the results of their meetings.

- *Gather information about the problem.* When the problems have been prioritized, the temptation will be to jump right in and begin solving them. This can be a mistake. The better approach is to collect all available information about a problem before pursuing solutions. Two kinds of information can be collected: objective and subjective. *Objective information* is factual. *Subjective information* is open to interpretation. Rarely will the information collected be only objective in nature. Nothing is wrong with collecting subjective information, as long as the following rules of thumb are adhered to for both objective and subjective information:

  - Collect only information that pertains to the problem in question.
  - Be thorough (it's better to have too much information than too little).
  - Don't waste time re-collecting information that is already on file.
  - Allow sufficient time for thorough information collection, but set a definite time limit.

Specifying the Problem

Specifying the problem means breaking it down into its component parts. All problems can be broken into five basic components as follows:

1. Who is the problem affecting?
2. What is the problem? This is a restatement of the problem specification.
3. Where did the problem occur first, when did it occur first, and when did it occur last? How often is it occurring?
4. Where does the problem occur? The answer to this question should be specific.
5. How much? What is the extent of the problem? This question should be answered in quantifiable terms whenever possible.

Identifying Causes

Identifying causes is a critical step in the process. An effective way to identify causes is the "Why-Why" exercise. Beginning with the problem specification, conduct a brainstorming session using the "Why-Why" method. It works like this: you state the problem and then ask "Why?" Participants brainstorm and propose ideas. Each idea is recorded. For each recorded idea you ask "Why" again. That "why" is answered and you ask "Why" again recording answers as you go. You keep asking "why" until you have run through symptoms and found the cause.

Isolating the Root Cause

The causes identified may or may not be the specific root cause of the problem in question. To isolate the most probable root cause or causes, each cause identified by the "Why-Why" activity is compared against the problem specification developed earlier.

When comparing potential causes and the problem specification for each cause, you will have three possibilities: The cause will fully explain the problem specification, the cause will partially explain the specification, or the cause will not explain the specification. A cause that can fully explain the problem specification is a likely candidate to be the root cause. If it takes more than one cause to fully explain the specification, there may be contributing causes of the problem.

Finding the Optimum Solution

With the problem and its most probable root cause identified, the next step is to find the optimum solution. The first task in this step is to develop a solution definition that clearly explains the effect the solution is to have. A solution definition should be the opposite of the problem definition. For example:

- Problem definition: Job ABC is over budget by 15 percent.
- Solution definition: Job ABC must be brought under budget.

With the solution definition in place, your team brainstorms possible solutions and develops a list. Perry Johnson, Inc., has developed a tool known as SCAMPER that can improve your team's effectiveness in building a solutions list. It is explained as follows:[2]

- *Substituting.*   Can the problem be solved by substitution? Can a new process be substituted for the old? Can one employee be substituted for another? Can a new material be substituted for the old?
- *Combining.*   Can the problem be solved by combining two or more tasks, processes, activities, operations, or other elements?
- *Adapting.*   Can the problem be solved by adapting an employee, a process, a product, or some other element to another purpose?
- *Modifying.*   Can the problem be solved by modifying a process, job description, design, or something else?
- *Putting to other uses.*   Can the problem be solved by putting a resource to other uses?
- *Eliminating.*   Can the problem be solved by eliminating a position, part, process, machine, product, or service?
- *Replacing.*   Can the problem be solved by replacing an employee, part, process, machine, product, or something else?

Note that the term *machine* is used generically in this model and includes any equipment or technology. With the list of potential solutions developed, the next step is to identify the optimum solution from among those on the list. One way to do this is by group consensus. A more objective approach is to conduct a cost-to-benefit analysis of each potential solution. Identify all costs associated with a given solution and record them with a total cost. Then list the benefits for each solution. The solution with the best benefit-to-cost ratio is the optimum solution.

SUCCESS TIP

The increasing pressures of a competitive marketplace are making it more important than ever for organizations to be flexible, innovative, and creative in decision making and problem solving. To survive in an unsure, rapidly changing marketplace, organizations must be able to adjust rapidly and change direction quickly. A reputation for being a creative problem solver—and for promoting creativity in others—will be a major benefit to your career advancement.

Implementing the Optimum Solution

The implementation phase of the process is critical. If handled properly, the problem will be solved in a way that results in permanent improvements. However, if implementation is handled poorly, new and even more serious problems can be created. In other words, solving one problem just creates other problems and nothing is gained.

The key to effectively implementing a solution is to take a systematic approach. Develop an implementation plan with the following components:

- Tasks to be completed
- Methods for completing each task
- Resources needed for each task
- Special needs relating to each task
- Person responsible for each task
- Deadline for completing each task

CREATIVITY IN PROBLEM SOLVING AND DECISION MAKING

The increasing pressures of a competitive marketplace are making it more important than ever for organizations to be flexible, innovative, and creative in decision making and problem solving. To survive in an unsure, rapidly changing marketplace, organizations must be able to adjust rapidly and change direction quickly. To do so requires creativity at all levels of the organization. A reputation for being a creative problem solver—and for promoting creativity in others—will be a major benefit in your career advancement.

Creativity Defined

Creativity has many definitions, and viewpoints vary about whether creative people are born or made. For the purpose of your career development, creativity can be viewed as an approach to problem solving and decision making that is imaginative, original, and innovative. Keep those factors—imagination, originality, and innovation—in mind when making decisions and solving problems.

Creative Process

According to H. Von Oech, the creative process has four stages: preparation, incubation, insight, and verification.[3] What takes place in each of these stages follows:

1. *Preparation* involves learning, gaining experience, and collecting/storing information in a given field or discipline. Creative decision making and problem solving require that the people involved be prepared. For example, you would not expect a nonengineer to solve a complex engineering problem.

2. *Incubation* involves giving ideas time to develop, change, grow, and solidify. Ideas incubate best when people get away from the issue in question and give their minds time to sort things out. Incubation is often a function of the subconscious mind.

3. *Insight* follows incubation. This is the stage when the light bulb turns on. It is the point in time when an idea for a solution finally falls in place and becomes clear. This point is sometimes seen as being a moment of inspiration. However, inspiration rarely occurs without having been preceded by perspiration, preparation, and incubation.

4. *Verification* involves reviewing the idea to determine if it will actually work. At this point, traditional processes such as feasibility studies, cost–benefit analyses, and engineering calculations come into play.

Factors That Inhibit Creativity

A number of factors can inhibit creativity in people. Some of the more prominent of these are as follows:[4]

- *Looking for just one right answer.*   Seldom is there just one right solution to a problem. More often there will be several potential solutions, each with its own good and bad points.

- *Focusing too intently on being logical.*   Creative solutions sometimes defy logic and conventional wisdom. The best time for logic is in the verification stage.

- *Avoiding ambiguity.*   Ambiguity is a normal part of the creative process. This is why the incubation step is so important.

- *Avoiding risk.*   When organizations don't seem to be able to find a solution to a problem it often means that key decision makers aren't willing to try a new idea.

- *Forgetting how to play.*   Adults sometimes become so serious they forget how to play. Playful activity can stimulate creative ideas.

- *Fear of rejection or looking foolish.*   Nobody likes to look foolish or feel rejected. Fear of rejection can cause people to hold back what might be a creative solution. Encourage open, nonjudgmental discourse when discussing problems in your team.

- *Saying "I'm not creative."*   People who decide they are not creative won't be. Any person can think creatively and learn to be even more creative. Make sure people you lead understand this.

Helping People Think Creatively

In the age of global competition, creativity in decision making and problem solving is critical. Although it is true that some people are naturally more creative than others, it is also true that any person can learn to think more creatively. In the modern workplace the more people think creatively, the better. Darrell W. Ray and Barbara L. Wiley recommend the following strategies for helping others think creatively:[5]

- *Idea vending.* This is a facilitation strategy. It involves reviewing literature in the field in question and compiling files of ideas contained there. Periodically, circulate these ideas among employees as a way to get people thinking. This will facilitate the development of new ideas by the employees. Such an approach is sometimes called *stirring the pot.*
- *Listening.* One of the factors that can cause good ideas to fall by the wayside is poor listening. People who are perpetually too rushed to listen to the ideas of others not only inhibit creativity, but stifle it. In addition to listening to the ideas, good and bad, of people, you should listen to the problems others discuss in the workplace. Every problem is grist for the creativity mill.
- *Idea attribution.* You can promote creative thinking in others by subtly feeding pieces of ideas to them and encouraging them to develop the idea more fully. When somebody develops a creative idea, he should receive recognition for the idea. Time may be required before this strategy pays off, but with patience and persistence it can help people become creative thinkers.

CRITICAL THINKING[6]

The most successful people are good critical thinkers. They have learned that things are not always what they appear to be on the surface, advice is not always good, and information is not always accurate. They know that failing to think critically can lead to mistakes, the adoption of ineffective strategies, bad decisions, and unnecessary disputes. Consequently, successful people understand that when solving problems and making decisions, critical thinking can be the difference between winning and losing.

Inaccurate assumptions, biased input, and bad information can cause you to make decisions that not only fail to solve the problem in question or produce the desired result, but actually make the problem worse. Critical thinking involves applying sound reasoning when making decisions and solving problems.

Behaviors of Critical Thinkers

We know that critical thinkers apply sound reasoning in all situations, but what does applying sound reasoning actually mean? People who apply sound reasoning do the following:

- Recognize bias in advice, recommendations, explanations, and information.
- Evaluate the motives of people who give advice, offer explanations, make recommendations, or provide information.

- Distinguish between facts and opinions.
- Distinguish between explanations and rationalizations.
- Separate real issues from inconsequential input.
- Take a 360-degree view of all problems and issues.
- Identify the fundamental issue at the heart of a problem.
- Diligently research the facts relating to issues (do their homework).
- Open-mindedly consider all potential alternatives, options, and solutions.
- Face difficult problems head-on and persevere in finding the optimum solution.
- Distinguish between the long-term solution and the short-term expedient.
- Question the thoughts, actions, and motives of people—including themselves.
- Use facts to patiently chip away at ambiguity.

Behaviors of Noncritical Thinkers

You will no doubt work with people who are not critical thinkers. I call such people *noncritical thinkers*. It is important that you recognize when 1) you are dealing with noncritical thinkers, and 2) you are exhibiting noncritical thinking behaviors yourself. If members of your team are noncritical thinkers, you must be prepared to point this fact out and show them how to think critically. If you find yourself falling into the non-critical thinking trap, you need to recognize the fact and make a quick adjustment.

Critical thinking is not a gift given at birth. Rather, it is a skill that must be learned, developed, and practiced constantly. Even the best critical thinker can fall back into non-critical thinking behaviors. Those behaviors include the following:

- Being closed-minded, inflexible, and stubborn when discussing issues and problems
- Being too receptive and willing to accept any opinion
- Being arrogant and overly confident about your own opinions
- Letting ego get in the way (e.g., it's my idea so it must be right)
- Vacillating between the opinions of others (e.g., accepting the latest opinion given)
- Reacting to problems and issues out of emotion rather than sound reasoning
- Unwillingness to get into the details of an issue
- Failing to question or consider motives, opinions, and the so-called *facts*
- "Piggybacking" (i.e., accepting another person's opinion without giving it due consideration)

Overcoming the Faulty Reasoning of Others

Critical thinkers often find themselves having to deal with the faulty reasoning of noncritical thinkers. What follows are some common manifestations of faulty reasoning and strategies for overcoming them.

Introducing Irrelevant Information

When discussing problems, noncritical thinkers often introduce irrelevant information. For example, assume your team is discussing a customer complaint. One member of your team—a noncritical thinker—says, "Who cares what this customer thinks about our engineering work? He doesn't do enough business with our company to matter." Sound reasoning would show that ignoring this customer's complaint, no matter how much business is at risk, would be a mistake. As a critical thinker, you would want to point this out.

How much business a customer does with your company is irrelevant. An unhappy customer who accounts for only a small amount of business can still spread the details of his dissatisfaction among other potential customers. In other words, what is at risk in this situation is much more than the small amount of business the unhappy customer in question accounts for. In addition, the company's goal should be to increase the amount of business received from every customer. This goal cannot be accomplished by ignoring customer complaints.

Ignoring Uncomfortable Facts

Noncritical thinkers often just ignore uncomfortable facts—facts that run counter to their preconceived notions. Assume you are leading an ad hoc team that is discussing various solutions to the problem of customer complaints about your company's computerized telephone answering service. Customers think the system is cumbersome, time-consuming, and impersonal. One of your team members says, "I don't care what customers say, I still think it's a good system." As a critical thinker, you would want to point out that this individual is ignoring some important facts. No matter how comfortable he might be with the answering system, if it causes problems with customers, it's not a good system.

Oversimplifying

Noncritical thinkers are often noncritical because they do not wish to invest the necessary time and effort to get to the heart of a problem. Consequently, they tend to oversimplify when dealing with problems. For example, assume that you and several other supervisors are discussing morale problems among employees in your company. One of your colleagues says, "There is nothing wrong with employee morale that a ten percent raise wouldn't fix."

You know from experience that morale problems are typically too complex to be solved by making just one change; even if that change is a pay raise. As a critical thinker, you would want to point out other factors that can affect employee morale which might be part of the problem with employees at your company. You might also want to suggest that rather than make assumptions about the causes of employee-morale problems, your fellow supervisors should convene meetings of their teams and talk with employees about morale issues. In other words, don't assume—ask.

Arguing from Ignorance

Noncritical thinkers often argue from ignorance. For example, assume your team is discussing ways to improve the performance of a certain process. One of your team members argues as follows: "I think we should just keep doing what we've always done. I've never seen a better way." This team member is arguing from ignorance. As a critical thinker, you would want to point out that just because this team member has never seen a better way doesn't mean there isn't one.

Using the "This-Is-How-Everybody-Else-Does-It" Argument

Noncritical thinkers often propose solutions and make recommendations based on the fact that everybody else does something a given way. For example, assume that you and several other supervisors are discussing how to improve a certain work process. One of the supervisors says, "I think we should use the ABC method. That's what everybody else does." As a critical thinker, you would want to point out that just following the crowd seldom makes sense. First, the crowd might be wrong. Second, even if the crowd is right, what works for others might not work for us.

Presenting a False Cause

Noncritical thinkers often present a false cause for a problem. For example, assume your team is discussing a customer complaint that is frequently received by your company. The complaint is that the main gear on the kitchen blender your company manufactures lasts only three to six months. If the blender is used to crush ice, the gear breaks down even faster. One of your team members suggests the following cause: "I'll tell you what's wrong with our blender. It's the way we cut the teeth on the main gear."

As a critical thinker, you would know that more investigation is needed before making such a definitive statement of cause. The problem with the gear might be the way the teeth are cut, but there is no evidence to support such a claim. The problem could also be caused by the material the gear is made of. In fact, this is a more likely cause since customer complaints about the gear did not begin to be heard until your company changed the gear material from stainless steel to plastic. Often people who present a false cause have an axe to grind—they are motivated by something other than finding a solution. In such a case, the team member who presented the false cause might have been upset for some reason about how the teeth are cut on the main gear for the blender. Consequently, he saw customer complaints about the gear wearing out as an opportunity to air his dissatisfaction about how the gear teeth are cut.

Using Circular Reasoning

Noncritical thinkers often use circular reasoning—reasoning that is supported only by itself. For example, assume that your team is discussing ways to improve a certain procedure. One of your team members says, "We need to eliminate Steps 3 through 5." You ask this team member how he knows that eliminating Steps 3 through 5 will improve the procedure. "I just know it will, that's how." As a critical thinker, you would want to point out that this team member is engaging in circular reasoning. Sound reasoning demands a better justification than "I just know . . ."

Using the Ad Hominem Approach

Noncritical thinkers often use the ad hominem approach to discount the recommendations of others. This approach involves discrediting the person making the recommendation rather than the recommendation itself. For example, assume that you and several other managers are discussing ways to make budget cuts. One of the managers says, "John Mantel thinks we should freeze out-of-state travel and restrict in-state travel." Another manager responds to this comment by saying, "What does Mantel know about budgets? He's never taken even one accounting course." As a critical thinker, you would want to point out that Mantel's education is not the issue. The issue is budget cuts and Mantel's recommendation might or might not have merit, regardless of his educational background.

Distorting the Conclusion

Noncritical thinkers often distort the conclusions drawn by others and then attack the distorted conclusion. For example, assume that you are a member of your company's senior management team. At a weekly management meeting, the company's chief financial officer (CFO) recommends that receipts for meals and hotel rooms be submitted when engineering personnel submit their requests for reimbursement after making business trips. Hearing this, the vice president for engineering protests as follows: "You are saying my people are dishonest. My engineers are highly ethical professionals. They don't need to submit receipts. If they say a hotel room or a meal cost a certain amount, they can be trusted."

Although you might agree with the vice president for engineering, as a critical thinker you would know that she has drawn a distorted conclusion. No one has questioned the integrity of the engineering personnel. Keeping receipts is common practice in business and it is good business. Auditors will certainly want to see receipts for expenses related to business travel. Requiring receipts is good stewardship, and the company's engineers are expected to be good stewards of company resources.

Using the Slippery Slope Argument

Noncritical thinkers often use the slippery slope argument when discussing issues. The slippery slope argument suggests that a given choice will automatically result in an ever-worsening set of circumstances. It's the old give-them-an-inch-and-they-will-take-a-mile argument. For example, assume you are a member of your company's management team. One of your colleagues has asked for a one-time increase in his department's budget to cover some unanticipated costs. The company's CFO responds as follows: "I'm against this proposal. You need to live within the approved budget like everyone else. If we approve this request, every other department manager will make the same kind of request."

As a critical thinker, you would want to point out that the CFO is using the slippery slope argument. You would also point out that requests for budget reallocations should be considered individually and on their merits. Just because one manager's request is approved does not mean that every manager's request must be approved.

Using Inflammatory Language

Noncritical thinkers will sometimes divert attention from the issue at hand by using inflammatory language. For example, assume that you and the other members of your company's engineering team are discussing a recommendation to promote a team member named Joe. One of the engineers says, "No way! Joe's a loser." This is inflammatory language in that it brands Joe in a powerfully negative way without providing reasons, evidence, or a satisfactory explanation.

As a critical thinker, you would want to point out that the term *loser* makes a powerful statement without telling the group what it really needs to know. You might ask the engineer who made this comment to provide specific examples that demonstrate why he thinks Joe is a *loser*.

Using Intimidating Language

Noncritical thinkers will sometimes use intimidating language to pressure or scare people into making decisions favorable to them (the noncritical thinkers). For example, assume you are part of your company's management team and the team is discussing the possibility of firing a supervisor who, after repeated counseling and mentoring, is still not performing acceptably. All members of the management team except one agree that this individual should be fired. The one who disagrees says, "You'll be sorry if you fire this supervisor. His brother is a lawyer. We'll be facing a lawsuit before the week is out."

As a critical thinker, you would want to point out that the decision to retain or fire the supervisor in question should be based on an objective assessment of his performance. The potential threat of a lawsuit simply suggests that the termination should be handled properly from a legal perspective. The fact that a lawsuit might result from the termination does not mean that an employee whose performance is consistently unacceptable should be retained.

Appealing to Compassion

Noncritical thinkers will sometimes appeal to compassion to influence a decision in their favor. For example, assume you are the leader of a team of engineers and technicians, and one of the employees on your team has been recommended for a salary increase by his supervisor. In studying the results of this employee's performance appraisals for the last two years, you determine that his work has been acceptable, but just barely. In truth, the employee in question is really just a marginal employee. When you point out the facts about this employee's lackluster work, his supervisor responds by saying, "I know his work isn't very good, but have a heart. He has a sick child in the hospital and the bills are really piling up."

As a critical thinker, you would want to point out that sound reasoning requires that salary increases be based on exemplary performance, not adverse personal circumstances. A more appropriate course of action for helping the employee in question might be to initiate a companywide fund-raising campaign.

Using Ridicule

Noncritical thinkers will sometimes resort to ridicule to embarrass others into making decisions favorable to them. For example, assume you are the engineering department manager and one of your supervisors is trying to persuade you to make a certain decision. The course of action you think best is different from the one the supervisor is advocating, and you tell him so. The supervisor then responds by saying, "You are going to look ridiculous if you make that decision. People are going to think you've lost your mind."

As a critical thinker, you would want to ask the supervisor to explain in detail how making a decision that serves the best interests of the company is going to look ridiculous. If the supervisor cannot make his case objectively and support it with facts, he has no case.

REVIEW QUESTIONS

1. List and explain the steps in the Perry-Johnson method of problem solving.
2. What is meant by specifying the problem?
3. Explain how you might go about identifying the causes of a problem.
4. Once you have identified potential causes of problems, how can you isolate root causes?
5. Explain Perry-Johnson's SCAMPER method for finding optimum solutions.
6. Use your own words to define the concept of *creativity*.
7. List and explain the stages in the creative process.
8. List at least five factors that can inhibit creativity in problem solving.
9. List five common behaviors of noncritical thinkers.
10. List and explain five strategies for overcoming the faulty reasoning of others.

SUCCESS PROFILE

Carol A. Sanchez
Engineer

Carol A. Sanchez serves as senior industrial engineer at Hughes' Missile Systems Group in Tucson, Arizona, where she provides production support to manufacturing teams that build her group's main product—missiles. By any standard, Sanchez has built a winning career in engineering. She has earned the Superior Performance Award, Hughes Aircraft Company's highest individual award, and the National Hispanic Engineer Achievement Award as the "Most Promising
(continued)

(*continued*)

Engineer." Sanchez completed her Bachelor's Degree in Systems Engineering from the University of Arizona.

Sanchez is the personification of the critical thinker, problem solver, and change agent. As an industrial engineer, these concepts are fundamental to her job. Continual improvement of productivity is her daily challenge and responsibility. In order to continually improve the performance of processes and people, Sanchez must think critically and she must constantly solve new and different problems. Every improvement she is able to make means change. Consequently, Sanchez must also be a positive change agent. As such she stresses the importance of communication and mentoring.

Continual improvement is about more than just identifying ways to make processes perform better. Processes are operated by people, and people cannot be reprogrammed as easily as computers and machines. Consequently, Sanchez stresses that engineers must learn not just the hard skills of their profession, but also the soft skills—the people skills. Communication, listening, respect for the opinions of others, persuasion, and consensus building are just as important to getting her job done as are the more traditional engineering skills.

Sanchez is an outstanding role model not only for young students who might consider a career in engineering, but also for those who choose to pursue a career in engineering. She participates in pre-engineering workshops conducted at the University of Arizona, lectures at local high schools, speaks at the annual Youth Convention of the League of United Latin American Citizens (LULAC), is co-chair of the Education Committee for the Hughes Hispanic Employees Association (HHEA), and helps operate a mentoring program to support Hispanic students at Tucson's Cholla High School.

The career of Carol Sanchez is one worth emulating. She has used and continues to use many of the success strategies recommended in this book in building her winning career. In particular, her career exemplifies the concepts of *problem solving*, *critical thinking*, *communication*, *change agency*, and *mentoring*.

Source: "Carol A. Sanchez." *Dictionary of Hispanic Biography.* Gale Research, 1996. Reproduced in *Biography Resource Center,* Farmington Hills, Michigan: Thomson Gale, 2005. *http://galenet.galegroup.com* (document number: K1611000384).

DISCUSSION QUESTIONS

1. Think of a problem you have had in the past. Now discuss how you could apply the Perry-Johnson method to solving that problem.

2. Defend or refute the following statement: Leave me out of this discussion—I am just not creative.

3. Assume you are having a discussion with a fellow college student who has asked for your recommendation in selecting from among several professors who all teach the same class. You have just recommended Professor Smith as the one you think will help students learn the most. Your friend says, "I don't care what you say, I'm not

taking a class from Professor Smith. I hear he is too hard." How would you respond to your friend's statement?

4. You are having a discussion with a fellow student who is thinking about downloading a term paper from the Internet and turning it in as if he wrote it. When you recommend against this practice, your friend says, "What's the big deal, everybody else does it?" How would you respond to this statement?

5. You are debating with a fellow student in class as part of an assignment. This student has made a claim that sounds questionable to you. In response, you say, "How do you know that approach will work?" Your opponent responds, "I just know it will." How would you respond to this statement?

ENDNOTES

1. Perry L. Johnson, Rob Kantner, and Marcia A. Kikora, *TQM Team-Building and Problem-Solving* (Southfield, MI: Perry Johnson, Inc., 1990), 1-1–10-15.
2. Ibid., 11-2.
3. H. Von Oech, *A Whack on the Side of the Head* (New York: Warner, 1983), 77.
4. Ibid.
5. Darrell W. Ray and Barbara L. Wiley, "How to Generate New Ideas," *Management for the 90s: A Special Report from Supervisory Management* (Saranac Lake, NY: AMACOM American Management Association, 1991), 6–7.
6. Daniel A. Feldman, *Critical Thinking* (Menlo Park, CA: Crisp Learning, 2002), 3–24.

CHAPTER SEVENTEEN

Learn How To Manage Conflict and Deal Effectively With Difficult People

Anger is just one letter short of danger.
    Anonymous

The most successful people in the workplace are good at managing conflict and at dealing with difficult people. This is an important *success skill.* Today's hectic, competitive, stressful workplace is a virtual factory for conflict. People in organizations have different agendas, ambitions, opinions, and perspectives. These differences and all of the other ways that people at work can differ contain the seeds of conflict. Add two other ingredients—ego and self-interest—and you have a potentially volatile mix. Without actually verbalizing it, many people subconsciously think, *"If you disagree with me, you must be wrong,"* or, worse yet, *"If you disagree with me you must be bad."*

Because so many people tend to see others who differ with them as being either wrong or bad, conflict in organizations is common. If this counterproductive situation is allowed to get out of hand, an organization can get so bogged down in negativity and conflict that it cannot perform at the level needed to be competitive. I've seen organizations with work environments that have so much conflict they could be described as being *toxic.* Learning how to prevent and resolve conflict is one of the keys to advancing your career.

In order to achieve consistent peak performance, organizations need their managers, supervisors, and employees working cooperatively toward achievement of the same goals. This does not mean they always have to agree—quite the contrary. Differences of opinion about how best to achieve the organization's goals can lead to better solutions, and better solutions are essential for continual improvement. This is the old philosophy that says you use fire to harden steel. Your opinion about the best way to do something might sharpen my opinion and vice versa—provided we are able to discuss our differing opinions intelligently and maturely without rancor, anger, and hurt feelings.

Performance cannot be continually improved in an environment where conflict has been allowed to turn negative and become counterproductive. In fact, in such situations the per-formance of an organization typically goes rapidly downhill. In order to improve continually,

SUCCESS TIP

In order to improve continually, organizations need to ensure the free flow of ideas. In order to ensure the free flow of ideas, organizations must have personnel who can disagree without being disagreeable.

organizations need to ensure the free flow of ideas. In order to ensure the free flow of ideas, organizations must have personnel who can disagree without being disagreeable.

Most successful people have learned how to manage conflict in organizations. This is one of the reasons they are successful. People who are good at conflict management know how to keep conflict on the high road that leads to better ideas and how to pull it up from the low road that leads to anger and strife. Marsha was such a person. She was the oldest in a family of ten children. Consequently, when she graduated from college and went to work, conflict management was nothing new to her. She had grown up refereeing arguments, breaking up fights, and resolving conflicts of all kinds among her younger brothers and sisters.

Marsha became especially good at helping her siblings see the futility in conflict. She showed them over and over how cooperating, sharing, and working things out among themselves was preferable to behaving in ways that would get "higher authority"—their parents—involved. When Marsha became an engineering supervisor, she used the skills she had developed at home to build consensus among her direct reports concerning how conflict would be handled in her team. Consequently, Marsha's team soon developed a reputation for unity of purpose. It became the *go-to* team for special projects and difficult assignments. As a result, Marsha's team members were frequently rewarded with higher percentages of incentive pay than other teams and Marsha advanced up the career ladder well ahead of her contemporaries.

In order to be effective at managing conflict and dealing with difficult people, you need to know how to do the following: (1) handle complaints; (2) turn complaints into improvements; (3) resolve conflicts between other people; (4) deal with people who bring their personal problems to work; and (5) deal with angry people. The remainder of this chapter explains strategies that can help you do these things.

SUCCESS TIP

Complaints that are viewed as opportunities and handled well can prevent minor problems from escalating into major problems, derail potential conflict situations before they get out of hand, lead to changes that result in improvements, and turn unhappy customers into loyal, repeat customers.

HANDLING COMPLAINTS

As soon as you have a position of authority in your organization, you will begin to receive complaints. All people in positions of authority—from the leader of the smallest team to the CEO of the largest organization—receive complaints from those they supervise, customers, and colleagues. Complaints that are viewed as opportunities and handled well can prevent minor problems from escalating into major problems, derail potential conflict situations before they get out of hand, lead to changes that result in improvements, and turn unhappy customers into loyal, repeat customers.

Complaints that are viewed as intrusions and ignored can lead to disaster. For example, many of the episodes of workplace violence that occur every year in the United States begin as minor complaints that fester and grow over time because they are consistently ignored by people in positions to do something about them. Internal complaints that are ignored or handled poorly can lead to quality, productivity, and morale problems. External complaints that are ignored or handled poorly can lead to lost customers and lawsuits.

When you secure a position of authority in your organization, no matter how busy you are, when an employee, colleague, or a customer contacts you with a complaint—*listen*. A good rule of thumb is to let all of your direct reports know they can have five minutes of uninterrupted time with you whenever they need it as long as they come prepared not just to complain, but also to recommend a solution. Once your direct reports know this, they will adopt a more positive attitude toward their work, their fellow employees, and you. It is also a good idea to thank customers who bring their complaints to you. After all, they could withhold their complaint and just take their business to one of your competitors.

Positive Listening Strategies

The most important skill for handling complaints, managing conflict, and dealing with difficult people is listening. If you can be a good, patient listener, you can also be an effective professional when it comes to handling complaints, resolving conflicts, and dealing with difficult people. When handling complaints, whether they come from employees, colleagues, or customers, apply the following strategies to improve your listening:

■ *Welcome the complaint as an opportunity to make improvements.*   As an engineering professional, you will be a busy person. Your days will be full even without the responsibility of dealing with people who want to complain. Consequently, the temptation to

SUCCESS TIP

The most important skill for handling complaints, managing conflict, and dealing with difficult people is listening. If you can be a good, patient listener, you can also be an effective professional when it comes to handling complaints, resolving conflicts, and dealing with difficult people.

treat complaints as unwelcome intrusions will always be there. Avoid this temptation no matter how involved you are with other pressing obligations. Professionals who ignore complaints because they are too busy are really saying, "I'm too busy to make improvements, retain a customer, prevent problems, or diffuse conflicts." People who are too busy to do these things will be lucky to keep the positions they have, much less advance in their careers.

■ *Eliminate all distractions.*   When an employee or a customer comes to your office to make a complaint, eliminate all distractions that might interrupt the meeting or distract your attention from what is being said. People with complaints are usually already frustrated. Don't add to their frustration by allowing telephone calls, drop-in visitors, or any thing else to interrupt their dialogue with you. Give people who wish to make a complaint your undivided attention.

■ *Sit up straight, adopt an interested expression, and maintain eye contact.*   I once worked with an engineer who had an interesting way of "listening" when people would bring complaints to him. He would sigh or groan loudly, put his head in his hands, and ask, "What is it this time?" It wasn't that he did this after spending a long and tiring day listening to a steady stream of complaints. He reacted this way upon receiving even the first complaint of the day. Needless to say, people were reluctant to bring complaints to this engineer, which might have been what he was trying to achieve in the first place. But if this was the case, his approach was a mistake—a mistake that eventually cost him his job when problems arose on an important project he was managing. These were problems he should have known about and would have known about if he had been more open to hearing complaints. Unfortunately for him and his company, because he made it so obvious that he didn't want to hear about problems or complaints, employees eventually got into the habit of bringing him only good news and either ignoring or covering up bad news. Even if you don't want to hear complaints from employees, colleagues, or customers, you may *need* to hear them. Consequently, when someone brings a complaint to you, sit up straight, put an expression on your face that says "I am listening, I am interested, and I care," look the speaker in the eye, and maintain eye contact throughout the discussion.

■ *Don't take notes.*   When someone brings you a complaint, there is a natural tendency to want to take notes. People who take notes when listening to complaints usually rationalize this practice in one of two ways: (I) "It shows the complainer I'm listening," or (2) "It gives me a record of the complaint so I don't forget it." Both of these reasons are based on laudable goals. You do want the complainer to know you are listening, and you do want to remember the basis of the complaint. However, there are several problems with taking notes that outweigh both of the reasons typically used to justify the practice. While you are writing down what is being said, you might miss something else—maybe something even more important than what you are writing down. While you are writing, you are not looking at the complainer. This can cause you to miss important nonverbal cues. Finally, the complainer is going to become frustrated if what you are writing down doesn't seem to match what he thinks are the important points. It's better to listen attentively while paying careful attention to nonverbal cues. Most complaints can be summarized mentally in just

SUCCESS TIP

Every person gives off a host of nonverbal cues when speaking. You can learn a great deal by "listening" with your eyes (i.e., watching for nonverbal cues). People communicate—often unintentionally—with their eyes, tone of voice, facial expressions, gestures, and posture.

one or two sentences. You can easily remember this much information. Mentally summarize the complaint. Then, after the complainer has left your office, write down your summary if you need to.

■ *Listen intuitively.* This strategy is sometimes referred to as "reading between the lines." Sometimes what the complainer doesn't say is just as important as what he does say. If you feel as if the complainer is leaving something out, stretching the point, shading the facts in his favor, or exaggerating, he probably is. Your intuition will tell you when these things are happening if you will listen to it. Later, I will explain what to do when your intuition tells you something is amiss.

■ *"Listen" with your eyes.* Every person gives off a host of nonverbal cues when speaking. You can learn a great deal by "listening" with your eyes (i.e., watching for nonverbal cues). People communicate—often unintentionally—with their eyes, tone of voice, facial expressions, gestures, and posture. If the complainer cannot look you in the eye or maintain steady eye contact, he might be lying to you or exaggerating. If the complainer's voice becomes shrill and quivers, he is probably nervous. You know how to read nonverbal cues. In fact, you could understand nonverbal communication before you could understand spoken or written communication. A little baby can tell if its mother is happy, sad, nervous, stressed, or angry even before it can understand the first spoken word. The baby can tell because its mother gives off nonverbal cues (e.g., how tense she is as she holds the baby and how her voice sounds as she talks to the baby). The key to understanding nonverbal communication is not, as many believe, to look for specific body language or gestures. Rather, it is to watch for agreement or disagreement between what is said verbally and what is "said" nonverbally. When you notice disagreement between the verbal and the nonverbal, ask the complainer for clarification. Inconsistencies in what the complainer says should be cleared up right away. Otherwise you might invest time and effort in solving a problem that doesn't really exist or you might solve the wrong problem. In the former case, you have wasted your time. In the latter case, you may have solved a problem, but it's not the one bothering the complainer. Consequently, you will still have the problem in question to deal with.

■ *Paraphrase the complaint and repeat it.* It is important for the complainer to know you have listened, and it is important for you to accurately understand the complaint being made. The best way to achieve both of these goals is to paraphrase the complaint and repeat it back to the complainer in your own words. For example, assume an employee

Success Tip

There are always at least two sides to a story. Never take a complaint you receive at face value. Before taking any action on a complaint, get the other side of the story—gather additional information.

has complained that he is expected to "work more than any one else on his team." He is clearly upset and has made negative comments about several of his teammates and the company. After his tirade has gone on for more than five minutes, you feel sure you understand the problem. It's now time for you to paraphrase the complaint and repeat it back to this employee. In this case, you might say, "Let me make sure I understand your complaint. You are upset because you've been asked to work overtime this weekend." By listening to what was said as well as what wasn't said, you were able to get to the heart of the problem and summarize it in just one sentence. His comments about being asked to work more than anyone else on the team were just hyperbole.

■ *Maintain your composure.* When someone makes a complaint, especially if they become emotional when making it, the natural human tendency is to respond in kind. If the complainer gets angry with you, a normal human response is to get angry with him. If the complainer becomes weepy, it's only natural for you to become weepy too. In both cases, you would be taking the wrong approach. When listening to complaints from customers, colleagues, employees, or anyone else in the work-place, you must maintain your composure. Adopt a nonjudgmental, reassuring atti-tude and maintain it. Think of yourself as being on stage in a play. Your role is that of the person who is going to receive the complaint and listen attentively to it. Your script says that you listen attentively and nonjudgmentally. It does not say that you lose your composure and respond in kind to the complainer. Play your role as it is written in the script—don't ad lib. When you have achieved a position of sufficient authority to receive complaints, if you lose your composure, you lose—period.

Using Complaints to Make Improvements

If you learn to handle complaints well, they can become an invaluable source of ideas for making improvements to your organization's performance. This is important because continually improving an organization's performance is one of the keys to success in today's competitive workplace. The following six-step model will help you use complaints to make continual improvements in your areas of responsibility:

1. Listen to the complaint.
2. Gather additional information.
3. Consider the broader implications of the complaint.
4. Take the appropriate action.
5. Monitor the results of your action.
6. Follow up with the person who brought the complaint.

Listen to the Complaint

The first step in the model involves listening to complaints in the ways explained in the previous section. Listen attentively, give the complainer your undivided attention, listen intuitively for what is not being said, watch for nonverbal cues, paraphrase and repeat the complaint back, and maintain your professional bearing regardless of the complainer's demeanor and attitude. Remember this is not just a complaint you are hearing, it is a potential opportunity to improve your organization's performance in some way, and improving your organization's performance is one of the ways you advance your career.

Gather Additional Information

There are always at least two sides to a story. Never take a complaint you receive at face value. Before taking any action on a complaint, get the other side of the story—gather additional information. In the previous section, you read the example of the employee who complained about being asked to work overtime on the weekend. Before taking any action on this complaint, you would want to know more about the situation. Why is overtime being required in the first place? Is the overtime covered in your department's budget? Are there, in fact, other employees who should be asked to work the overtime instead of the one who complained? How often is the employee in question asked to work overtime as compared with other team members? Before taking any action, find the answers to these questions and any others that might be relevant. This will do two things for you—both of them positive. First, it will give you the information you need to undertake the next step in the model—*consider the broader implications.* Second, it will help you make a better decision concerning the resolution of the complaint.

Consider the Broader Implications of the Complaint

In the previous step in the model, you gathered information that will be useful in this step. Information concerning why the overtime was required in the first place and the budget ramifications of the overtime relate to the broader implications of the complaint. In this step you consider these and any other potential implications that are broader or more important than the actual complaint itself. For example, I once received a complaint from an employee that his paycheck seemed to be short by around $100. To this employee, the complaint related only to his individual pay. But to me the complaint had much broader implications. If this employee's check was off by $100, every employee's check might be off by that amount or more. Paychecks for my organization were generated by a computerized payroll system. Chances were good that if one check was incorrectly calculated, all checks were. After listening to the employee's complaint, I asked my organization's payroll department to look into the issue right away. Sure enough, every check printed on that day was short by some amount. People who work for a living are serious about their paychecks. Even small errors can cause big problems. Because we caught the computer glitch in time to quickly inform all employees of the problem, I had to deal with only one angry employee rather than the entire workforce. The original complaint was about one paycheck, but it had broader implications that went well beyond just one employee.

Take the Appropriate Action

Once you have listened to a complaint, gathered additional information about it, and considered its broader implications, it's time to take the appropriate action. Sometimes the most appropriate action is no action. Not all complaints are valid. But assuming the complaint is valid, you now take whatever action is necessary to solve the problem—both the specific problem of the individual who brought the complaint and any broader problems you have determined to exist. In the paycheck example used previously, the appropriate action was to (1) Correct the paycheck of the individual who complained and thank him; (2) Let all employees know that their checks would be short by some amount due to a computer glitch, but that the situation was being corrected immediately and they would get their full amounts; and (3) Examine the computerized payroll system to determine what had gone wrong with the payroll calculations and correct the problem.

Monitor the Results of Your Action

Don't ever assume that your action has solved a problem. Once you have taken what you think is the appropriate action, monitor the results to make sure the solution is working. If you correctly identified the root cause of the problem and took the action necessary to eliminate the root cause, your solution will work. However, sometimes what we think is the root cause of a problem is just a symptom. If your solution treats symptoms rather than the root cause, the problem will recur. Consequently, it's important to monitor the results of your action just in case you need to try another solution.

Follow Up with the Person Who Brought the Complaint

One of the most common complaints from employees and customers is that once they complain about a problem, they never hear what happened. People need you to close the loop for them. When people complain, they want to know what, if anything, happened. Was the problem solved? Will it happen again? Did my complaint do any good? People don't like to be left hanging. They want to know what happens as a result of their complaints even if what happens is nothing.

Consequently, once you have taken whatever action you think is appropriate and monitored the results to make sure the solution adopted is working, follow up with the person who brought the complaint. Let that person know you took action, monitored

SUCCESS TIP

When you have employees involved in conflict, don't wait to see what might happen—take action right away. Bring the employees together in a private setting; give them an opportunity to state their grievances without interruption, contradiction, or judgmental comments from you or each other.

the results of the action, and solved the problem. If you decided to take no action, let the complainer know that and why. Closing the loop on complaints lets people know they have been taken seriously and that their complaint has been given due consideration.

Resolving Conflicts between Other People

Once you achieve the position of team leader, supervisor, or higher, you will need to be effective at resolving conflict. When counterproductive conflict occurs between employees who are your direct reports, it can quickly spread as other employees choose sides in the battle. Consequently, it is important to get counterproductive conflict resolved quickly.

When you have employees involved in conflict, don't wait to see what might happen—take action right away. Bring the employees together in a private setting; give them an opportunity to state their grievances without interruption, contradiction, or judgmental comments from you or each other. Treat the employees with respect and let them know you take their conflict seriously. The message they need to get from you is this: "You are both important to the team. I need you to get past this conflict and work together cooperatively for the good of the team. Let's talk about how that can happen."

Once you have let each party to the conflict state his case without interruption or contradiction and let each know that the goal of the meeting is resolution and cooperation, apply the following strategies to move them toward resolution and cooperation: (1) help participants identify the source of the conflict, (2) let participants know you expect all issues to be discussed in a mature and positive manner, (3) let participants know they are responsible for resolving the conflict, (4) let participants propose solutions and then discuss their proposals, and (5) guide participants toward a positive resolution.

Help Participants Identify the Source of the Conflict

People in conflict tend to attribute the conflict to the personal motives and shortcomings of the other party. In fact, there are several predictable factors that are often the root cause of many workplace conflicts. These factors include the following: communication problems, differing points of view, insufficient resources, counterproductive territoriality, and differing agendas. It is important to help employees see that their conflict is based—more often than not—on logical, understandable factors rather than on malice, bad intentions, revenge, jealousy, power, greed, ego, ignorance, or other counterproductive human issues.

Communication problems play a part in most workplace conflict. Even when poor communication is not the root cause of the problem, it is usually a contributing cause. Having listened to both parties state their positions, does it appear that poor communication is contributing to the conflict? "You and I disagree because there has been insufficient communication" is a much less explosive, easier-to-accept cause of conflict than the causes people tend to attribute to others who disagree with them (e.g., malice, greed, power, ego, ignorance). If poor communication appears to be a contributing factor, point this fact out to both parties.

People can have widely differing points of view concerning the same issue. For example, think of a candidate in a presidential election. Many people will think he is wonderful while others will think he's the worst person who ever ran for office. How can people see the same individual so differently? It's simple—they have different perspectives. Some people are liberal while others are conservative. Some people are extroverts while others are introverts. Some people are big-picture oriented while others see only the details. People have different points of view. When someone has a different point of view than you, the tendency is to ascribe the difference to something bad or wrong in the individual. You will have to help employees in conflict realize that having a different point of view does not make a person bad. It doesn't even make him wrong.

Expect a Mature and Positive Discussion of the Conflict

At this point in the process, I typically give each party in the conflict a copy of the organization's mission statement. I explain that the mission statement represents the reason for our employment, and that our job is to help the organization accomplish its mission. Personal agendas have no place in the discussion. I explain that I've brought the two parties together to have a mature and positive discussion about issues, and that everything said during the discussion should be viewed in the context of how it helps the organization accomplish its mission.

This is usually an effective way to say, "It's not about you—it's about us." Often, when employees get this message the conflict evaporates and nothing further need be done. However, in those cases when the discussion does need to continue, you have at least converted the context from petty bickering to a mature discussion.

Let Participants Know They Are Responsible for Resolving the Conflict

Often when team members bring their disagreements to you, they are trying to do a *hand-off*. In other words, they are trying to hand off their problem to you and make you solve it. All this will accomplish is the transfer of blame from the conflicting parties to you. Don't let this happen. Your role is more that of the referee than the judge in this situation. Let participants know that they are responsible for resolving the conflict. Your role is to facilitate the process.

SUCCESS TIP

Often when team members bring their disagreements to you, they are trying to do a *hand-off*. In other words, they are trying to hand off their problem to you and make you solve it. All this will accomplish is the transfer of blame from the conflicting parties to you. Don't let this happen.

Ask Participants to Propose Solutions

The best conflict solutions are those that come from the participants themselves. Once you have helped the parties to the conflict identify its source, established the ground rules for a mature, positive discussion, and reminded participants of their responsibility in resolving the conflict, ask them one at a time to propose solutions. Hear both parties out and don't allow interruptions, comments, or judgmental behavior from one party when the other is proposing a solution. Your main task in this step is to get proposed solutions on the table so that the context of the discussion switches from conflict to solutions.

Guide Participants Toward a Solution

Your main task in this step is to guide the participants toward the optimum solution. An effective way to do this is to use the organization's mission statement as the basis for conducting a quick cost–benefit analysis of proposed solutions. The best solution is the one that most effectively serves the organization's mission. Don't make the mistake of trying to find a compromise solution that will please both parties. The most popular solution is not always the best solution from a business perspective.

DEALING WITH PERSONAL PROBLEMS AT WORK

Many organizations tell employees to leave their personal problems at home when they come to work. This is management's way of telling employees they don't want the company's performance to be hindered by the personal problems of employees. Although the principle is understandable, just telling employees to leave their personal problems at home is no guarantee they will, or even that they can. This is one of the reasons it is so important for you to maintain an appropriate balance between the work you are responsible for and the employees who do that work.

When team members or direct reports experience personal problems that affect their work, your job is to be a supervisor or a colleague, not an amateur psychologist. The key words in the previous sentence are "... *that affect their work.*" Unless an employee's work performance is hindered by his personal problems, those problems are none of your business. What makes them your business is adverse effects on the employee's work.

When an employee's personal problems begin to adversely affect his performance at work, document the work-related problems and monitor the situation for a short period of time. If you are patient for a few days, the situation might clear itself up and the employee's work performance might right itself without intervention. However, there is an important caveat to this principle. If you see evidence that the employee's downturn at work is related to drinking or drugs, you must act immediately. Go directly to your organization's Human Resources Department and let them know about your suspicions. On the one hand, you must never accuse an employee of abusing alcohol or drugs, no matter how sure you are this is happening. On the other hand, you must never let an impaired employee conduct any work-related tasks in which the impairment could lead to an injury or the destruction of property. Whenever an employee's abilities seem to be impaired, get the Human Resources professionals in your company involved immediately and do not allow the employee to perform any work tasks until given a release by the Human Resources Department.

When you notice a downturn in an employee's work, simply ask to talk with him. Explain the problems you have observed and documented. Focus on the evidence you have of declining performance and keep the discussion on a professional level. The best approach is to explain why you think there is a problem, solicit feedback from the employee, and listen. Remember, you will learn more from listening than from talking. Once the employee has stated his case, ask him to propose a solution, and remember, even if the employee brings up his personal problems, you should stick to job performance issues. Don't get pulled into a discussion of an employee's personal problems. While personal problems might make declining work performance understandable, they do not make it acceptable. Work with the employee to develop a plan for improving performance. Set specific goals and time frames. Do not try to be an armchair psychologist. Be a supervisor, team leader, manager, or colleague.

DEALING WITH ANGRY PEOPLE AT WORK

Your career advancement prospects will improve markedly when you learn how to deal effectively with angry people. Whether the angry person is a customer, colleague, direct report, or team member, you will help your career by learning to do three things: (1) stay calm as you deal with angry people; (2) take the steps necessary to calm the anger of other people; and (3) transition angry people from a negative state of mind (anger mode) to a positive state of mind (solution mode).

Staying Calm When Confronted by an Angry Person

When dealing with angry people, do you respond by fighting fire with fire? Some people do. They respond by returning the anger they receive. Although understandable, this is the worst response you can take. The most successful people have learned the following rule of thumb: *When dealing with an angry person, if you lose your temper, you lose— period.* When you respond to an angry person with anger of your own, you have started down a path that can lead to a bad place. Anger that is answered with anger is likely to just escalate until it gets out of control.

Try the following strategies to stay calm when confronted with anger. First, take a few deep breaths to settle your breathing and keep it normal. Second, ignore the anger and listen in between the lines for the substance of the problem. The more angry people become, the less able they are to state clearly what's bothering them. They tend to exaggerate, leave out facts, add in superfluous facts, and generally make more noise than sense.

The key to staying calm in such situations is to learn to ignore all of the *irrelevant noise* generated by the anger while focusing instead on listening for useful facts. When conducting seminars on this topic, I use the following example from the medical profession to explain the concept of ignoring the noise of anger while listening, instead,

SUCCESS TIP

> This is how it is when dealing with an angry person. If you get hung up focusing on the anger, you will miss the source of the problem. Take a couple of deep breaths, ignore the anger, and listen for the source of the problem.

for its substance. A man gets a splinter in his forearm and cannot get it out. After a couple of days his arm becomes infected. The area around the splinter becomes swollen, red, and extremely painful—so painful that the man decides to go to the emergency room of the local hospital. At the hospital, the emergency room physician looks at the man's arm to determine the problem. He has two choices: (1) he can treat the swelling, redness, and infection around the splinter; or (2) he can ignore the swelling, redness, and infection, find the splinter, and remove it. If he chooses the first option, the splinter will still be in the man's arm and the infection will return. If he chooses the second option, the arm will begin to heal the moment the splinter is removed. This is because the first option treated only the symptoms of the problem while the second option eliminated its source.

This is how it is when dealing with an angry person. If you get hung up focusing on the anger, you will miss the source of the problem. Take a couple of deep breaths, ignore the anger, and listen for the source of the problem. The harsh words, exaggerations, and even threats associated with anger are like the swelling, redness, and infection in the man's arm from the example above—they are symptoms, not causes. If the physician in this example forgot to stay calm and let himself get distracted by the swelling, redness, and infection in his patient's arm, he might never have located and removed the source of the problem.

Take Steps to Calm the Angry Person

If you can stay calm when confronted with anger, you can begin taking the steps necessary to calm the person in question. What follows are several strategies you can use to help calm an angry person:

■ *Listen and let the angry person vent without interruption.*   Angry people are like teakettles. If not given the opportunity to vent, they might explode. Consequently, one of the most effective strategies for dealing with an angry person is to simply sit back, listen to them without interruption, and let them vent. While listening, look directly at the person who is venting and give him an affirming expression that says, "I'm listening and I care about what you have to say." Avoid nonverbal cues such as shaking your head in disagreement, moving away defensively, crossing your arms, rolling your eyes, or making negative facial gestures. Just listen in a nonjudgmental frame of mind and with body language that conveys that frame of mind.

■ *Acknowledge the anger.*   Don't interrupt while the angry person is venting, but when his tirade seems to have run its course, acknowledge the anger. You can do this by saying something as simple as: "You're really angry about this, aren't you?" or "I can see you are really angry about this." Most people, having been given an opportunity to vent without interruption or contradiction and having had their anger acknowledged, will calm down. If this doesn't happen, continue to listen and acknowledge the anger.

■ *Use a simple apology as a bridge.*   When dealing with angry people, the goal is to help them make the transition from the anger mode to the solution mode. A well-crafted apology can be an effective bridge between these two modes. *Well-crafted* means that the apology is brief, direct, and solution-oriented. It does not go on too long or become maudlin. An example of a well-crafted apology is as follows: "I am sorry this happened. Let's see what you and I can do to correct the problem." Notice that immediately after making a very brief and direct apology, you transition to the solution mode. You also enlist the formerly angry person as a partner in solving the problem.

■ *Paraphrase and repeat back.*   Once you have transitioned the person in question from the anger mode, let him begin to propose solutions. Once again listen attentively and do not interrupt. After each solution that is proposed, paraphrase and repeat it back in your own words. This will let the person in question know you have listened. It will also give him an opportunity to correct your perception if it is slightly off.

■ *Ask open-ended questions to clarify and to solicit additional information.*   If what the Person in question proposes is not clear or is not well thought out, use open-ended questions to clarify and to solicit additional information. An open-ended question cannot be answered "yes" or "no." Such questions begin with the following types of phrases: "Tell me about . . . ," "What do you think about . . . ," or, "What are your thoughts on . . ." You can also use open-ended questions to guide the person you are talking with through a mental cost–benefit analysis when his proposed solution is unrealistic or poorly formulated.

■ *Confirm the eventual solution.*   Once a realistic solution has been arrived at, confirm it with the person in question. Do not assume that you understand both the solution and all of its ramifications without first confirming it. A good solution is one that, once put in place, will stay in place. It eliminates the source of the problem that caused the anger in the first place or it reveals to the formerly angry person that the problem was caused by a miscommunication or misunderstanding, as is often the case.

REVIEW QUESTIONS

1. What are the five main things you need to know how to do in order to manage conflict and deal with difficult people?
2. Explain how you should handle complaints made by employees or customers.
3. Describe how you can use complaints to make improvements.
4. Explain the process for helping other people resolve conflicts they are having with each other.

5. Explain how you should handle the situation when the work of employees who report to you is being negatively affected by their personal problems.
6. Explain the various strategies for dealing with angry people at work.

DISCUSSION QUESTIONS

1. Discuss how you would handle the following situation: An employee who reports to you stops you in the hallway and complains that she can't get her work done on time because the other members of her team spend too much time on coffee breaks.
2. Discuss how you would handle the following situation: Two of your team members are constantly bickering with each other. Their bickering is beginning to bother other team members to the point that their work is suffering. Both employees are good at their jobs and you would like to keep them in your team.
3. Discuss how you would handle the following situation: Mary, who is one of your best team members, is having problems with her husband. You have met Mary's husband several times, so you sympathize with her. Her personal problems are beginning to undermine the quality of Mary's work.
4. Discuss how you would handle the following situation: One of your team members stomps into your office, slams the door, and yells, "I hate this stupid place. Management never gives us the support we need to do our jobs!"

CHAPTER EIGHTEEN

Be a Mentor to Team Members and Your Boss

It can be argued that in a democratic society, people have a birthright: to become all they can be. And mentors help their mentees move toward fulfilling that birthright.[1]

Gordon F. Shea

When trying to build a winning career, an excellent way to help yourself is to help others. One of the ways successful people become successful is by helping others they work with develop useful knowledge and skills; knowledge and skills that enhance the performance of the team and the overall organization. This concept is known as *mentoring*.

Mentoring is typically associated with a more experienced, more senior person helping with the development of a promising, but less experienced, less senior person. I call this type of mentoring *pull mentoring* because it involves reaching down in the organization and pulling promising employees up. The rationale for pull mentoring, from your perspective, is that the better your team members perform, the better your team will perform. And the better your team performs, the better your career will develop.

There is another type of mentoring that gets less attention and is used less frequently but is just as important as pull mentoring. I call this other approach *push mentoring*. Push mentoring involves building up your boss—using your knowledge and skills to help him in ways that enhance his performance. The rationale for push mentoring is that one of the best ways for you to climb higher on the career ladder is to help the person on the rung above you make room by climbing higher himself.

PULL MENTORING

Pull mentoring is the most widely used form of mentoring. Gordon F. Shea describes this approach to mentoring as follows: "A developmental, caring, sharing, and helping relationship where one person invests time, know-how, and effort in enhancing another person's growth, knowledge, and skills, and responds to critical needs in the life of that person in ways that prepare the individual for greater productivity or achievement in the future . . . anyone who has a beneficial life- or style-altering effect on another person, generally as a result of personal one-on-one contact; one who offers knowledge, insight, perspective, or wisdom that is helpful to another person in a relationship which goes beyond duty or obligation."[2]

I first learned about pull mentoring shortly after finishing college and beginning my career. I worked in a team of engineering professionals led by a person who had more experience than all of his team members combined. I will call this person John Smith. It was the good fortune of all of us in his team that this thoroughly experienced professional was willing to share his considerable knowledge and wisdom with us. Twice weekly John Smith would hold what he called *chalk talks*. For an hour or so after work, he would meet in a conference room equipped with a chalk board and flip charts and help his direct reports develop knowledge and skills that would prove invaluable not just to the performance of our team, but to the development of our careers.

Smith seemed to know what it took to succeed in our profession, and he had an uncanny ability for recognizing the individual strengths and weaknesses of his team members within the context of what was necessary to succeed. Every time we met in that conference room, he would have at least one specific recommendation for improvement for every participant. Sometimes he would make assignments and we would have to report our progress in completing the assignments at the beginning of the next meeting.

One Christmas season our company's CEO decided to host a formal Christmas banquet and invite all professional-level personnel to attend—although *invite* is not exactly the right word here. Attendance was mandatory. The younger engineering professionals on John Smith's team were anxious to make a good impression on the CEO as well as the rest of the company's senior managers. Consequently, we were concerned about such things as how to dress and proper table manners. None of us owned a tuxedo and not even one of us could distinguish a salad fork from a dessert fork or a water glass from a wine glass. We had all just spent four years in college where our most formal mode of dress had been jeans and where good table manners were optional. Manners aren't an issue when all you eat is hot dogs, hamburgers, and pizza in the college's student center, your dormitory, or a fraternity house. My colleagues and I were worried about making a bad impression on senior managers by demonstrating conclusively our total ignorance of business etiquette.

John Smith had sensed our anxiety about the upcoming Christmas banquet. Consequently, at his next mentoring meeting, he came prepared. For three meetings in a row, he brought in a friend who was a business etiquette expert. During these meetings, my colleagues and I learned how to properly wear a tuxedo or a formal gown, and the best places in town to buy or rent them. We also learned about proper cocktail conversation, formal place settings, table manners, and proper table talk.

Thanks to John Smith, when the big day arrived for the CEO's Christmas banquet, my colleagues and I were ready. We were well prepared and even confident. In fact, as things turned out, we all made positive impressions on the CEO and the entire senior management team.

This is just one example of the many ways John Smith helped develop the inexperienced engineering professionals on his team. He also helped us develop winning attitudes, think critically, make appropriate choices, and develop critical skills. Smith's mentoring not only helped us, but also helped the company and, in the long run, it helped him. Because of John Smith, my colleagues and I performed at a much higher level than we would have without his help. We knew this, and that knowledge generated a strong sense of loyalty in us toward Smith.

SUCCESS TIP

In order to create a win-win-win mentoring situation, you need to understand the responsibilities of mentors as well as what the most effective mentors do. Generally speaking, in order to be effective, mentors must be willing to give of their time, remain open-minded, be able to give feedback that is constructive and tactful, and listen well.

Out of a sense of gratitude and loyalty, my colleagues and I worked extra hard to give John Smith the highest performing team in the company—and we succeeded. Smith repeated his practice of mentoring his direct reports in each successively higher position he attained. Because of the consistent peak performance of his teams, Smith rose steadily through the ranks in our company and enjoyed a very successful career. He helped us, himself, and his company by helping his team members. As you can see from this example, mentoring—when done right—can be a win-win-win enterprise.

In order to create a win-win-win mentoring situation, you need to understand the responsibilities of mentors as well as what the most effective mentors do. Generally speaking, in order to be effective, mentors must be willing to give of their time, remain open-minded, be able to give feedback that is constructive and tactful, and listen well. In addition, there are specific responsibilities mentors must be willing to accept. The next section explains those responsibilities.

Principal Responsibilities of the Mentor

The most effective mentors accept responsibility for undertaking the following tasks on behalf of those they are trying to help:

- Communicating openly, frankly, tactfully, and frequently.
- Serving as a sounding board and patient, attentive listener.
- Providing encouragement, recognition, and support.
- Providing a steady flow of accurate, up-to-date information about opportunities, issues, problems, and options.
- Being a consistent role model of success-oriented attitudes and behaviors.
- Helping set goals and realistic timetables for their achievement.
- Helping develop effective strategies for achieving goals.
- Helping understand the organization's culture and how to operate successfully in it.
- Making introductions to useful contacts.
- Helping to develop specific job- and career-related knowledge and skills.

Push Mentoring

Push mentoring is upward mentoring. It's about making your boss better. This aspect of mentoring gets very little attention in professional literature, but it's important nonetheless. Too many engineering professionals just out of college think that in order to move up the career ladder they need to knock their boss off of it. This is a mistake—taking the low road to career advancement always is. It might appear to work out in the short run, but in the long run you will pay a price for taking this approach.

Recall the example of John Smith from the previous section on pull mentoring. Smith had many professional strengths that served him well as he rose through the ranks of our company. But he also had some weaknesses. During our biweekly mentoring meetings, Smith would sometimes reveal certain of his weak areas to us and ask for help. Of course, those of us he was helping would practically line up to return his favors. The fact that he was willing to ask for help increased his credibility with us markedly, and—as a side benefit—bolstered our confidence.

Your boss might not be as open to revealing weaknesses as John Smith was, but even if he isn't, there will be weaknesses and if you are attentive you can recognize them. Here are some ways you can help your boss: (1) be a second pair of eyes and ears for him and share what you see and hear; (2) speak positively about him; (3) work quietly to improve his image and performance; and (4) point out potholes along the way before he steps into one.

Be a Second Pair of Eyes and Ears

Nobody can be everywhere at once. Your boss, no matter how bright, is not omniscient. Help him out. If there is a controversy brewing that could affect his team's performance, let him know about it. If there is a problem in the team he is unaware of or does not accurately perceive, talk to him. If you learn that someone above your boss has issues with him or the team, let him know. Anything you hear, see, or learn about that could potentially help or harm your boss, pass it on to him. If you have recommendations for dealing with the problem or issue, ask if he would like to hear them. If so, make your recommendations. It is true that your boss will see and hear things you won't. After all, his is a higher branch on the tree than yours. But it is equally true that direct reports hear and see things their boss doesn't. When this happens, talk to your boss. Keep him informed.

Success Tip

Nobody can be everywhere at once. Your boss, no matter how bright, is not omniscient. Help him out. If there is a controversy brewing that could affect his team's performance, let him know about it. If there is a problem in the team he is unaware of or does not accurately perceive, talk to him. If you learn that someone above your boss has issues with him or the team, let him know.

Speak Positively About Your Boss

Too many people in the workplace think they gain some type of benefit by running down their boss in conversations outside of his hearing or by participating in gossip about him. Both of these activities are mistakes. First, this type of thing almost always finds its way back to your boss. Second, you will lose credibility with those who hear your negative comments—even those who appear to appreciate them. Always speak positively about your boss in public and private. If you have issues with your boss, take them up with him privately in a face-to-face conversation. This approach will help build the relationship while, at the same time, showing your boss he is dealing with a loyal team member.

Work Quietly to Improve Your Boss' Performance

Every boss has his strengths and weaknesses. When you observe an area of weakness in your boss, work quietly to strengthen him in that area, or try to mitigate the effects of the weakness in other ways. Whatever you do to improve your boss's performance, do it quietly and behind the scenes. You don't help your boss by publicly pointing out discrepancies. This might make you look smart, but it will make your boss look bad—never a good idea—and it might drive a wedge between you and your boss no matter how well intended your actions were. Remember this unchanging rule of career development: *Getting at odds with your boss will not help your career.*

I got an opportunity to help improve my boss' performance in my first professional position after graduating from college. I had a knack for writing and a boss who didn't. His letters to customers and memorandums to colleagues suffered from poor grammar, sentence structure, and spelling. Don't get me wrong. This individual was bright—he just wasn't a writer. After reading only a couple of his memorandums, I knew that his limited writing skills were going to hurt his image, if they hadn't already. This individual was a good boss and I wanted to help him, but the situation presented me with a real dilemma. How do you tell your boss his writing is terrible—especially when you have just graduated from college and had your new job for just six months?

Rather than confront him with a memorandum that I had marked up and corrected, I decided to wait for an opportune opening. One day while I was in his office, he made a comment about how much he hated to write. I saw his statement as an opening and immediately dashed into it. I said, "Most people hate writing, but I'm one of those strange people who actually likes it. Why don't you just draft your letters

SUCCESS TIP

Too many people in the workplace think they gain some type of benefit by running down their boss in conversations outside of his hearing or by participating in gossip about him. Both of these activities are mistakes. First, this type of thing almost always finds its way back to your boss. Second, you will lose credibility with those who hear your negative comments—even those who appear to appreciate them.

and memorandums and let me take care of the details? I'll just mark them up for grammar and sentence structure. Then your secretary can clean up the draft. It's a waste of your time to spend it laboring over the details of grammar and syntax." My boss liked the idea and handed me a draft letter on the spot. The little time it took me to uphold my end of this partnership was negligible but important—it helped both of us. Nobody but my boss and I knew about the arrangement. Everything was done quietly. Even his secretary knew only that he had begun to give her marked-up drafts of letters and memorandums that she would correct. My boss returned this minor favor many times over and eventually made a significant contribution to my success.

At another point in my career I worked for a boss who was often invited to speak to civic clubs and other community-based organizations. He liked public speaking and did it passably well so long as the podium met certain requirements. If it didn't, he would become noticeably uncomfortable, and his discomfort would affect his speech. After watching him in action several times, I was able to identify the cause of his inconsistent performance. He had an interesting quirk. Because of a vision problem, he needed a podium of just the right height and with plenty of light. When he had the right podium, he was a fairly effective public speaker. However, if the podium did not meet his requirements, the quality of his speech would typically range somewhere between disjointed and disastrous.

When I figured out what was causing my boss' inconsistent performances, I started serving as a front man for him. I asked my boss to let me handle the logistics whenever he was invited to speak and he agreed. From that point on, whenever my boss received an invitation to speak, I would arrive at the designated location early and check things out. If the podium wasn't just right, I would try to make appropriate adjustments. I might put books under the podium to raise the height, or move it to another location to increase the amount of light shining on it. However, some podiums just don't lend themselves to adjustments, and getting the necessary amount of lighting was always tricky.

After struggling with these problems several times with only mixed results, I decided to find a permanent solution. After looking through numerous catalogues—the Internet had not been invented at this time—I found a portable podium with its own battery-operated light that was height adjustable. The perfect solution! I convinced my boss to order this innovative product. When the portable podium arrived in the mail, it looked like a cross between a small suitcase and a large briefcase. Its height could be easily adjusted by simply turning a knob on the side of the podium, and it even contained several extra batteries and bulbs for the light.

SUCCESS TIP

Everybody makes mistakes—even the most talented, experienced, and intuitive people. Like you and everyone else, your boss will make mistakes. However, you should do everything possible to keep your boss from making mistakes that might harm his career, the performance of his team, or the image of the organization.

After my boss had given a couple of public speeches using what he called his *wonder podium,* I never had to serve as front man for him again, and he never had to worry about podium problems again. The quality of his speeches improved and the number of speaking invitations he received increased—both of which helped his career and, as a result, mine.

Point Out the Potholes Before Your Boss Steps into One

Everybody makes mistakes—even the most talented, experienced, and intuitive people. Like you and everyone else, your boss will make mistakes. However, you should do everything possible to keep your boss from making mistakes that might harm his career, the performance of his team, or the image of the organization. I learned this lesson years ago when a subordinate of mine kept me from making what could have been a serious mistake. I was preparing for a meeting in which I expected my department to have its budget cut substantially. Our executive vice president had warned all department-level managers to expect budget cuts during a particularly difficult economic period for our company. This vice president had come up through the ranks from the field of marketing, and he made no secret of his preference for marketing over the company's other functional units.

Consequently, I was preparing an impassioned defense of my department in which I planned to really challenge the comparative value of marketing. I was going to use graphs, charts, and slides to prove that my department was more important than all of the other departments put together—especially marketing. My thinking went like this: By the time I was done with my presentation, every department except mine would have its budget cut. Mine, on the other hand, would get a well-deserved increase.

I asked several members of my team to join me in the conference room for a trial run of my presentation. After I had finished practicing my impassioned and self-righteous appeal, one of my team members spoke up and said the presentation was right on target and that I should wade in with both guns blazing. Most of the team members in the room agreed. One seemed to capture the feelings of the group when he said, "Hit 'em hard. It's about time somebody told our vice president there is more to this company than marketing."

Then, amid all the ensuing trash talking and bravado, a quiet voice from the back of the room said, "I don't think that's a good idea." When I asked this team member to expound, he told me that my approach might feel good at the moment, but that in the long run it would just create powerful enemies for our department. He explained that it would be better to work on convincing all department heads, including the one from marketing, to be united in approaching the vice president with a proposal for an equal distribution of the budget cuts on a pro rata basis. In this way, rather than making enemies, we would be building partnerships with the other departments that might serve us well; and not just in the current situation, but in the future too.

Although I still wanted to make a fight of it, I saw the wisdom in this team member's input—so I took his advice and, as it turned out, I was glad I did. Not only were the other department heads willing to present a united front to the vice president, but the marketing manager actually made unity a moot point when, before anyone else could speak, he offered to take a larger share of the cuts in his department. Shocking everyone, our pro-marketing vice president accepted the marketing manager's magnanimous offer.

I had originally intended to take the floor first and preempt any action by our vice president with my impassioned presentation of fiscal reality as I saw it. Had I done this, not only my action would have insulted a colleague—the marketing manager—who had come to the meeting prepared to be selfless and equitable, but it would have turned the meeting into a battleground. Thanks to the wisdom and sound advice of one of my direct reports, I got through the meeting without (1) making a complete fool of myself, and (2) making a powerful enemy. As an added bonus, my department as well as those of my colleagues received less of a budget cut than they otherwise would have. My team member kept me—his boss—from making a serious mistake.

Keep your eyes and ears open and use what you learn to help your boss. Talk him up, not down. Help improve or mitigate any weaknesses he might have, and never let him make a mistake that will harm his career, his image, or the performance of his team. Quietly take care of your boss in these ways and your boss will take care of you. Of course, there will be the occasional exception to this rule. Not all bosses are ethical, equitable people. But this principle will apply favorably in the vast majority of cases.

REVIEW QUESTIONS

1. Explain the term *pull mentoring.*
2. What are the principal responsibilities of the mentor?
3. Define the term *push mentoring.*
4. What is meant by being a second pair of eyes and ears for your boss?
5. Why is it important to always speak positively about your boss?
6. Explain how improving your boss' performance can enhance your career.
7. Why is it important to keep your boss from making mistakes?

DISCUSSION QUESTIONS

1. Discuss how helping others enhance their work performance can help you enhance your career.
2. Discuss how you should handle the following situation: You are at a company social function and a colleague says, "I hear your boss is really a tyrant."
3. Assume your boss is disorganized and his disorganization causes problems that are beginning to be noticed. Discuss how your boss' problems could hurt your career and how you might help your boss in this situation.
4. How would you handle the following situation: Your boss has gone to great trouble and effort to arrange a dinner party for an important client your company needs to impress. You learn quite by accident that this particular client hates dinner parties.

ENDNOTES

1. Gordon F. Shea, *Mentoring* (New York: AMACOM, American Management Association, 1994), 43.
2. Ibid., 13–14.

CHAPTER NINETEEN

Learn to Balance Work and the Rest of Your Life

I learned about the strength you can get from a close family. I learned to keep going, even in bad times. I learned not to despair, even when my world was falling apart. I learned that there are no free lunches. And I learned the value of hard work. In the end you've got to be productive. That's what made this country great—and that's what's going to make us great again.[1]

Lee Iacocca

In his book, *Today Matters*, John Maxwell says, "People are funny. When they are young, they will spend their health to get wealth. Later they will gladly pay all they have trying to get their health back."[2] There is wisdom in what Maxwell says. When you are just starting a career, the temptation is to put everything else aside while charging full speed ahead doing what is necessary to climb the career ladder. For many people, it isn't until later when they have wrecked their health, ruined their relationships, and lost their families that the wisdom of Maxwell's words begins to sink in.

While it is true that with age comes perspective, it is equally true that you don't have to wait. You can gain an appropriate perspective right now if you choose to. Remember that smart people learn from their mistakes, but wise people learn from the mistakes of others. My hope is that in reading this book, you will choose to be wise and seek to learn from the mistakes made by others who have pursued their careers to the detriment of all else. The subjects of the study that led to this book overwhelmingly advocated an approach to career development that can best be summarized as follows: *Learn to balance work with the rest of your life.*

The following story is illustrative of this point. At twenty-two years of age, Ben was a fresh college graduate with a new job in his field and a promising career ahead of him. At the age of twenty-six he was a rising star in his company. At thirty Ben was a department-level manager. At thirty-five he was a corporate vice president—the youngest in his company's history. At forty Ben had risen to the pinnacle of his profession and was the CEO of his company. His alma mater named him a distinguished alumnus, the local chamber of commerce selected him as its business leader of the year, and he was featured in local newspapers.

Shortly after becoming a CEO, Ben's meteoric rise came to a screeching halt and his good fortune suddenly went bad. At fifty years of age, Ben was divorced, friendless, and

SUCCESS TIP

The workplace you will enter after college is full of people who have built careers by sacrificing their families, friends, and health. This situation is unfortunate from any perspective, but it is especially unfortunate when you consider that a supportive family—whether it consists of mother, father, sisters, brothers, wife, husband, children, uncles, aunts, grandparents, other significant people, or any combination of these—can be your greatest asset in building a winning career.

suffering from health problems including high blood pressure and heart disease. Ben had focused so intently on developing his career that he completely neglected his family, friends, and health—a mistake that cost him dearly.

Ben's story is all too common in today's intensely competitive, persistently demanding workplace. Too many people are like Ben in that they pursue success in their careers so single-mindedly that they neglect everything else in their lives. The successful people interviewed during the study that led to this book were almost unanimous in their conviction that the so-called *success* bought at the cost of family, relationships, and health is not success at all. Rather, they associated success with maintaining an appropriate balance between career and the other important aspects of life (e.g., family, relationships, and health).

The purpose of this chapter is to encourage you to take the long view in pursuing career success and to maintain an appropriate balance in your life—a balance that will allow you to enjoy the success you achieve and to share it with people who care about you and for whom your success matters.

You have to decide for yourself what the important elements of life are beyond your career. For most of the people interviewed during the study that led to this book, these elements included family, friends, and health. There were other elements such as faith and service mentioned by some of the study subjects, but family, friends, and health were the elements mentioned most frequently. Consequently, this chapter provides recommendations for balancing your career with family, friends, and health.

BALANCING FAMILY AND CAREER

Family relationships are foundational to your life. More than anyone else, family members will be there to cheer when you succeed and console when you don't—and even the most successful career will be full of both successes and setbacks. But family relationships, in order to be supportive, are like a machine—they require regular maintenance. Family maintenance consists of time, communication, attention, and caring. Give your family members these things, and they will give you the support needed to build a winning career. Then, when you get to the top in your profession, you'll have people who care about you and who will be there to share your success.

The workplace you will enter after college is full of people who have built careers by sacrificing their families, friends, and health. This situation is unfortunate from any perspective, but it is especially unfortunate when you consider that a supportive family—whether it consists of mother, father, sisters, brothers, wife, husband, children, uncles, aunts, grandparents, other significant people, or any combination of these—can be your greatest asset in building a winning career.

What follows in the remainder of this section are strategies that will help you balance your career obligations and family responsibilities. If you do not yet have a spouse or children, you can still apply these strategies to your other family members (mother, father, sisters, brothers, aunts, uncles, and others who are significant to you).

Convene Periodic Family Meetings and Communicate Openly

As soon as you finish college and secure a position in your field, convene a family meeting. Have an open two-way conversation with family members in which you explain your career ambitions, expectations, and goals. A good way to do this is to share the latest version of your career plan (Chapter 4) with family members. Let family members know you will need their support in order to achieve your career goals, and that you would like your career development to be a family project. As best you can do this early in the process, let family members know what they can do to help you advance up the career ladder.

Once you have set your career goals before them and asked for their help, ask family members to voice their thoughts or concerns. Ask them to help you identify the times and events in their lives when your presence is especially important to them. Talk about ways you and your family members can stay in touch on those occasions when your work will require out-of-town travel.

Ask family members to discuss the issue of balancing career and family obligations. Get their thoughts on how best to achieve the goal of balance. Try to identify specific things you and family members can do for each other to help achieve an appropriate balance. To the extent possible at this point in your career, try to paint a word picture of how daily family life will be as you pursue your career.

It is important to minimize to the extent possible any career-related surprises to your family. As you begin applying the strategies in this book to advance your career, try to anticipate how career obligations will affect family members and let them know as far in advance as possible.

Over time, hold family meetings regularly. In these meetings, have an open, two-way conversation about progress and problems. Let family members know how your career development is progressing and how specific things they do are helpful. If anything they are doing or failing to do is hindering your career development, tell them. But do so in a tactful and caring way.

Give family members ample opportunities to voice concerns and problems about the issue of balance. Listen attentively to what family members say, and do so in a nonthreatening, nonjudgmental manner. Open, continual communication is essential to achieving the goal of balance between family and work responsibilities. But the effort you put into communicating with family members will be worth it. In the long run, a family partnership will take you farther in your career than you could go working alone.

Success Tip

To make sure you spend an appropriate amount of time with your family, and to make sure you don't forget important family events, block out family time on your calendar in the same way you block out work-related appointments.

Schedule Family Time

The most successful people have learned to be good time managers. They use a calendar—either electronic or written—as a time-management tool. The same time-management strategies that serve you well at work will serve you equally well at home. To make sure you spend an appropriate amount of time with your family, and to make sure you don't forget important family events, block out family time on your calendar in the same way you block out work-related appointments.

Then, once you have blocked out time on your calendar for family, guard those times. Don't get into the habit of treating scheduled family time as tentative appointments you keep only if nothing work-related comes up. Guard family time on your calendar as vigorously as you guard important work-related appointments.

Schedule Alone Time with Your Spouse

If you are married, be sure to schedule alone time with your spouse—and do so regularly. The marriage relationship requires regular maintenance. Uninterrupted talking time is an important part of that maintenance. Open, frequent communication can make your spouse your most effective career-development partner. On the other hand, a neglected spouse may come to resent your career. When this happens, a marriage-versus-career situation can develop that will cause you to lose on both accounts.

Remember Special Family Days and Events

During family meetings, identify the family days and events that are most important and put them on your calendar. I've talked with many people who now regret having missed family "firsts" because they were tied up with work responsibilities. At the time, the work obligations may have seemed more important, but time often brings a different perspective. I have a friend who cannot bear to look at family videos because he is rarely in them. He missed his son's first steps, first day at school, numerous baseball games, and even his high-school graduation. Now he cannot even remember what work-related obligations were so pressing that he chose to miss these important family events.

Help Family Members Achieve Their Goals

A partnership is a two-way, mutually beneficial enterprise. Consequently, if you want your family members to be partners in your career development, you will have to be

SUCCESS TIP

My recommendations for balancing a career and your health are as follows: (1) don't smoke, and avoid the smoke of those who do; (2) eat for nutrition, not comfort; (3) exercise regularly; (4) don't use drugs; (5) get plenty of rest; and (6) drink only moderately—if at all.

a partner in helping them achieve their goals. Talk with family members about their aspirations in life. Help them develop a plan like the career plan you developed in Chapter 4. Ask family members to identify specific ways in which you can help them achieve their goals. If you have a child who wants to make the baseball team at school, you might need to block out time each week to practice with him. If you have a child who wants to win a scholarship to college, you might need to block out time to help her study. If your spouse works, you should help him achieve his career goals. The best way to convince family members to be partners in your development is to be a partner in theirs.

BALANCING HEALTH AND CAREER

A college friend of mine outpaced me and the rest of our graduating class in advancing up the career ladder. He and I stayed in touch after graduation by telephone—this was in the ancient days before e-mail. My friend always ended our telephone conversations with the same line: "The next time you are in Tampa, come see me." It took years, but I finally paid my friend a visit. We met for supper and I soon learned that my friend's career was in great shape, but he wasn't. In fact, his health was a mess.

Since graduating from college, he had gained forty pounds and become a chain smoker. He also had high blood pressure and heart problems. His diet consisted primarily of black coffee and junk food. While my friend had worked to build his career, he had completely neglected his health. I am sad to say that this bright, ambitious, talented professional died a young man—an all too common occurrence in today's hectic workplace.

Although you might achieve some short-term career gains by neglecting your health, in the long run poor health habits will catch up with you. My academic preparation includes degrees in five different fields. However, I have no formal training in any academic discipline even relating to health. Consequently, my recommendations in this section are just the common-sense wisdom of a layperson enhanced by what I have learned from experience over the last forty years.

My recommendations for balancing a career and your health are as follows: (1) don't smoke, and avoid the smoke of those who do; (2) eat for nutrition, not comfort; (3) exercise regularly; (4) don't use drugs; (5) get plenty of rest; and (6) drink only moderately—if at all. What follows in the remainder of this section are more specific explanations of each of these health-related strategies.

Don't Smoke, and Avoid the Smoke of Those Who Do

This health strategy is simple enough. Smoking is bad for you—don't do it. Just read the warning written on the side of every cigarette package. It will tell you all you need to know about smoking and health. If you are smart enough to complete a college degree, you are too smart to smoke. This includes exposing yourself to the second-hand smoke of others.

Eat for Nutrition

When it comes to fuel, all too often we treat our cars better than we treat our bodies. Food is first and foremost fuel for the body. The main reason for eating is nutrition—to provide fuel for the body. You wouldn't pour valve-clogging sludge into the gas tank of your car, but many of us pour artery-clogging sludge in the form of fats and sugars into our bodies every time we eat.

When we eat for convenience, comfort, or any reason other than nutrition, we tend to eat too much and we eat the wrong things. The following rules of thumb will help you maintain a healthy diet:

- Balance your caloric intake so that you maintain a mix of carbohydrates, proteins, and fats that approximates the following ratio: fifty percent carbohydrates, thirty percent protein, and twenty percent fat.
- Get the bulk of your carbohydrates from the complex varieties (e.g., raw vegetables and fruit) while avoiding the simple carbohydrates found in refined foods and sweets.
- Avoid junk foods—they are high in both fats and sugars. Avoid trans fats.
- Drink plenty of water—eight or more glasses per day—while limiting your intake of other types of drinks; especially those high in caffeine.
- Eat whole-grain breads and rice rather than the more refined white varieties.
- Eat plenty of fiber (e.g., bran, raw vegetables, and fruits).
- When you eat meat, select the leanest varieties (e.g., white meat chicken, fish, and turkey). Make sure the meat is grilled or baked rather than fried, and avoid high-fat sauces and gravies.
- Limit your intake of alcoholic beverages, and if you drink—don't drive.

Even fast-food restaurants have healthy options if you look carefully and order wisely. Most serve salads and bottled water. When ordering a salad, be careful of the salad dressing you choose. Read the labels closely. You can negate the positive benefit of a good salad by using a bad dressing that is high in calories and fat. Many fast-food restaurants serve grilled chicken or sliced turkey sandwiches. If you order a chicken or turkey sandwich, make sure to ask for whole wheat bread and leave off the mayonnaise. Ask the server to pile your sandwich high with lettuce and tomatoes, and substitute mustard for mayonnaise.

If you are like me, it's going to be hard to give up pizza, doughnuts, sweets in general, and chips. I love these things, and the old saying "you can't eat just one" was written with me in mind. Actually you don't have to completely give these things up. If you apply all of the rules of thumb just presented about eighty percent of the time, you can get away with occasionally sneaking a piece of pizza and other goodies on special occasions; at least while you are still in your twenties. But remember this, the older you get the more important it will be to limit the amount of junk food and goodies that make their way into your diet. Once you reach thirty years of age and older, you will probably need to stick to the rules of thumb presented herein at least ninety percent of the time.

In fact, once you reach no more than thirty years of age, it is a good idea to begin having annual health checkups that include a complete blood workup. Make sure to ask the physician to conduct a comprehensive blood test (e.g., cholesterol, blood sugar, triglycerides).

Exercise Regularly

Regular exercise is one of the most beneficial things you can do for your health. The best exercise programs include both aerobic and resistance exercises. Aerobic exercises are those that tone up and maintain the circulatory system (e.g., heart, lungs, arteries, veins). Walking, jogging, biking, and various stationary exercises have high aerobic value, as do sports that require a lot of running such as basketball and tennis. Resistance exercise involves working out with weights or doing push-ups and pull-ups. These types of exercise tone and maintain the skeletal muscles and rev up your metabolism.

When working out, *consistency* is more important than intensity. Don't get started on an exercise program that is so intense and takes so long that you can get in a good workout only occasionally. Find a workout routine that involves both aerobics and resistance training you can realistically do regularly, including when you are traveling. For example, I typically do three resistance workouts and five aerobic workouts per week. All of these workouts are done Monday through Friday. In other words, on three days, I do both weight training and aerobics. My weight-training workout is a full-body routine that takes just forty-five minutes. My aerobic routine is a mix of stair-stepping, stationary biking, running, walking, and cross-country biking. My aerobic routine is never less than thirty minutes, but seldom more than forty-five minutes. Only when walking or cross-country biking do I work out for an hour or more—and then only as time permits.

My exercise routine will not put me in Arnold Schwarzenegger's league, but it does keep me in shape and I can do it consistently, even when on the road. Speaking of working out when traveling, it is a lot easier now than it was when I first graduated from college—so no excuses. In the early days of my career, getting in a good workout while on the road was a challenge. In those days there weren't fitness centers in most towns

SUCCESS TIP

The list of successful people who have ruined their lives and careers by abusing illegal drugs would fill a book larger than this one. Don't add your name to this list. You are trying to build a winning career. Illegal drugs are for losers.

SUCCESS TIP

When you begin drinking to bolster your courage, settle your nerves, or drown your sorrows, you have begun down a road that can lead to disaster. Drinking too much has ruined the careers and lives of many good and talented people. Don't become another one of them.

and hotels. I can remember getting lots of strange looks as I did pull-ups while hanging from the limbs of trees or handstand push-ups while leaning against the wall in the corridor of a hotel.

Things are different and much better now. In even the smallest town you can usually find a fitness center that will allow you to work out as a nonmember by paying a reasonable daily rate. In fact, many hotels and motels now have their own fitness centers. Regardless of the facilities available, do not make excuses for missing a workout while traveling. In small towns you can always do push-ups, sit-ups, and find a nice place to walk or run. In big cities, there will be fitness centers available for your resistance workouts, and for aerobics you can always take a brisk walk around the block several times or, better yet, do what I like to do—run the stairs. The really high multistory hotels in major cities afford you an opportunity to get in one of the most challenging aerobic workouts you'll ever endure. Find the stairwell used for emergency exits, go to the lowest floor you can access, and start running up. I like to run from the bottom floor of the hotel to the top, turn around, and walk back down, then repeat the process as many times as it takes to get in a good workout of thirty to forty-five minutes. If you run out of gas before getting to the top, you can always exit the stairwell on any floor and walk that floor until you catch your breath.

Don't Use Drugs

The list of successful people who have ruined their lives and careers by abusing illegal drugs would fill a book larger than this one. Don't add your name to this list. You are trying to build a winning career. Illegal drugs are for losers.

Get Plenty of Rest

Get plenty of rest. Creative thinking requires a clear head. This might turn out to be the most difficult to take of all the health-related advice offered in this section. You might have heard the adage, *"Don't work hard, work smart."* According to the successful people interviewed during the study that led to this book, this adage is misleading. It gives people the false impression that if they work smart they won't have to work hard. That's nonsense. While it is certainly true that successful people work smart, it is equally true that they work hard. They also work long. One of the ways successful people become successful is that they outthink and outwork the competition. In order to have a clear enough head to outthink the competition and enough energy to outwork them, you will need plenty of rest. Avoid the temptation to work all day and party all night.

This practice can catch up with you faster than you might think, and when it does *burnout* is the usual result.

If You Drink—Drink Moderately, and Don't Drive

For many, drinking is an ingrained part of social interaction. There is no problem with this as long as the drinking is done in moderation. However, drinking enough to inhibit your ability to interact intelligently in a social or business setting is not going to help your career. Drinking and driving could end it—permanently. What happens to some people who end up drinking too much is that they start out using alcohol as a social lubricant and wind up using it as self-medication. When you begin drinking to bolster your courage, settle your nerves, or drown your sorrows, you have begun down a road that can lead to disaster. Drinking too much has ruined the careers and lives of many good and talented people. Don't become another one of them.

BALANCING RELATIONSHIPS AND CAREER

Remember when you graduated from high school and went on to college. What happened to all of those relationships you had made in high school? If you are like most people, you have probably drifted away from high-school friends and made new ones in college. The same thing can happen when you graduate from college and begin building your career. Some of this drifting apart is to be expected as old friends are replaced by new ones. But remember this: Relationships are important.

As you begin building your career and find yourself drifting farther and farther away from old friends, ask yourself the following question: "If I were to die tomorrow, who are the people I would want to attend my funeral?" Make a list of these people and keep it somewhere private. Don't forget to include family members, but keep the list relatively short. The people on your list should be meaningful to you—much more than just acquaintances or partying buddies. Then, periodically pull the list out and update it. You will find that over time some names will drop off the list and others will be added. This is just a normal part of life and nothing to fret about.

At any given time, the people who remain on your list are people you should make a special effort to stay in touch with. These are people whose relationships with you matter. They should be people who pass the following test: *They know the real you, but love you anyway.* Depending on proximity, you can stay in touch in person—this is always best—or by telephone, e-mail, or old-fashioned letters. The best gift you can

SUCCESS TIP

> The best gift you can give a friend is time. Time invested in them is the best way to build, maintain, and nurture relationships with those who matter most to you.

give a friend is time. Time invested in them is the best way to build, maintain, and nurture relationships with those who matter most to you.

REVIEW QUESTIONS

1. Explain in your own words why it is important to balance family and career if you want to have a winning career in the long run.
2. Describe when and how to convene family meetings for discussing career-related issues.
3. Explain how scheduling family time can actually help your career in the long run.
4. Explain in your own words why it is important to balance health and career.
5. What is meant by "eating for nutrition"?
6. Describe an exercise routine that will help keep you healthy in the long run.
7. What is meant by *burnout* and how can it be avoided?
8. Explain the importance of balancing relationships and your career.

DISCUSSION QUESTIONS

1. Defend or refute the following statement: "I'll worry about health, family, and relationships later. Right now the best thing for me to do is go 24/7/365 on my career."
2. Discuss which of the following individuals you think is having the more successful life. Sarah has risen steadily up the career ladder in record time. At just thirty-five years of age Sarah is a CEO, wealthy, and an influential person. She is also divorced, beginning to experience health problems, and estranged from her only child. Margie, at forty years of age, is a department director with a good chance of becoming a vice president in her company within another five years. She is not wealthy, but her salary is respectable. She describes her financial condition as "comfortable." Margie is healthy and enjoys several close relationships with people who are important to her.
3. As he rushes up the career ladder, John is living off black coffee, cigarettes, and junk food. He is making excellent progress in his career, but the rest of his life is a mess. Assume John is your friend. If he asked for advice on how to get his life moving in a better direction, what would you tell him? Discuss your response with others.

ENDNOTES

1. Louis E. Boone, *Quotable Business*, 2nd ed. (New York: Random House, Inc., 1999), 267.
2. John C. Maxwell, *Today Matters* (New York: Warner Faith, 2004), 86.

CHAPTER TWENTY

Persevere Through Difficult Times—Don't Quit and Never Give Up

A winner will find a way to win. Winners take bad breaks and use them to drive themselves to be that much better. Quitters take bad breaks and use them as a reason to give up.[1]
Nancy Lopez

A career is like a tennis match—it consists of more than just one game. In a tennis match there are several games in a set and several sets in a match. Even the most talented tennis champions will lose a game here and there, but they are champions because they persevere through to ultimate victory in spite of these intermittent losses. They don't let the occasional failure cause them to become disheartened and give up. One of the uniform characteristics of successful people is that they have learned to view every setback as an opportunity for a comeback.

Perseverance will be essential to your long-term success in building a winning career. There will be times when you won't get the promotion you deserve, the job you want, the raise you need, or the recognition you should. Life can be good, but it's not always fair. It is when life deals you the inevitable setbacks that perseverance becomes so important, and there is no question that you will have setbacks. Things won't always go your way. When this happens, remember that every setback is an opportunity for a comeback. Understand that every failure comes to you with a gift in its hand. That gift is the opportunity to learn from the experience and try again—this time wiser and better prepared.

SUCCESS TIP

> Winning is not just about talent. The pathway to failure is paved with the unfulfilled dreams of talented people who, when faced with adversity and hard work, just gave up. They quit. Sometimes victory goes to the person who is willing to just hang in there the longest.

The subjects interviewed in the study that led to this book were unanimous and assertive in offering the following advice: (1) never quit and don't give up, (2) be fiercely persistent, (3) take adversity in stride and do what is necessary to overcome it, and (4) when times are hard just keep going. Strategies for applying these principles are presented in the remainder of this chapter.

Don't Quit—Never Give Up

Winning is not just about talent. The pathway to failure is paved with the unfulfilled dreams of talented people who, when faced with adversity and hard work, just gave up. They quit. Sometimes victory goes to the person who is willing to just hang in there the longest. I think of America's Olympic women's softball team that had to work so long and so hard just to get their sport into the Olympics, but persevered and were eventually successful. But they did not stop there. Having won the battle for Olympic acceptance, America's women then went on to win three consecutive gold medals in their sport. I think of the boxer who loses the first nine rounds of the fight, but perseveres until finally in the tenth round, he knocks out his opponent. I think of the times baseball games have been won when the last batter in the bottom of the ninth inning clubs a home run and of football games won by a field goal as time expires. I think of great basketball games won at the buzzer by a desperation shot tossed up from mid-court. In all of these examples, victory went to the individual or team that persevered, that refused to give up or quit. Sometimes victory is just one step beyond that last step you think you can take.

All of my life I have seen people either succeed or fail based on their willingness or unwillingness to hang in there just a little longer than the competition. I write here not of sports, but of career choices made in the workplace. Of course the best-known and most dramatic examples of victory being snatched from the jaws of defeat at the last minute come from the world of sports. But the never-quit-never-give-up philosophy applies just as directly in the workplace as in the athletic arena.

I observed an illustrative example of perseverance in action while still a college student. A classmate was pursuing a technical degree that required the entire Mathematics sequence from Algebra through Calculus. This classmate was smart, but not in the Mathematical sense. Academically, he was more comfortable in such disciplines as History, Political Science, and English. In fact, he had no desire to pursue a career in a technical field. Instead, he wanted to be an attorney. Learning of this, I asked him why he was pursuing a technical degree. Why not major in Political Science and then go on to law school?

His response showed just how determined a person he was. His family was poor and could not afford to help with the expenses of college. The only way he could afford college—this was in the days before financial aid was so readily available—was to accept a scholarship presented to him by an organization that stipulated he pursue a technical degree. Researching the matter thoroughly before accepting the scholarship, this classmate found that a good foundation in a technical field coupled with the law degree he planned to pursue would be excellent preparation for a career as a patent attorney. Armed with this knowledge, he accepted the scholarship and set

SUCCESS TIP

Each time you try and fail, you are better prepared for the next time. A failed attempt is not a failure unless you quit. A failed attempt is just another opportunity to try again better prepared for success this time.

himself on a course to be a patent attorney by way of first earning a bachelor's degree in a technical field.

Unfortunately, there was a major stumbling block between my classmate and the realization of his career goal. That stumbling block was Mathematics. His Math skills were not up to completing even the first course in the Math sequence required in his undergraduate degree program. Not one to be discouraged by roadblocks, he enrolled in college-preparatory Math classes at a nearby community college and continued taking them until he felt ready to begin the college-level Math sequence.

Even with what he had learned in his college preparatory work, Math was still a struggle. But my classmate refused to give up. I lost touch with this determined student after my graduation, but later learned that he did persevere and eventually graduate; but only after having repeated several college-level Mathematics courses he was required to take. I've often wondered if he got into law school and, if so, how he did. I don't know, but I do know this. I'll bet he excelled at law school and in his career as a patent attorney because the words *quit* and *give up* were not in his vocabulary.

Strategies for Persevering When You Want to Quit

Perseverance is more of a mental than a physical concept. In fact, I refer to perseverance as the mental equivalent of physical stamina. It's what lets you hang in there a little longer when what you really want to do is quit. Whenever you encounter difficulties in your career, apply the following strategies. They might help you keep going a little longer or pick yourself up and try again.

- Remember the lesson of the great inventor Thomas Alva Edison. In trying to invent such useful products as the light bulb and the storage battery, he failed repeatedly. It is said it took him almost 25,000 attempts to finally succeed. But he wouldn't quit. Edison refused to give up. He persevered, and finally succeeded. We should all be thankful he did. People who appear to be "overnight successes" typically aren't. In reality most have been working hard at succeeding for many years.

- Think about how hard you have worked to get where you are. Think about the advice my favorite football coach from the old days, Vince Lombardi, used to give his players during the golden years of the Green Bay Packers. Paraphrased, Lombardi's message to his players went like this: *The harder you work at accomplishing a goal, the harder it is to give up on achieving it.* In other words, don't quit now— you've worked too hard to get this far.

■ Each time you try and fail, you are better prepared for the next time. A failed attempt is not a failure unless you quit. A failed attempt is just another opportunity to try again better prepared for success this time.

■ Don't focus on what might happen if you fail—focus on what will happen when you win.

FACE ADVERSITY AND OVERCOME IT

Successful people are not strangers to adversity. My favorite example of facing adversity and overcoming it is Franklin Delano Roosevelt, president of the United States leading up to and during World War II. Roosevelt was elected president when the United States was still firmly in the grip of the Great Depression. Unemployment was at its highest level in the nation's history, the nation's banking system had crashed, small businesses were closing daily, people were losing their homes because they couldn't pay their mortgages, the Midwestern farming states had been turned into a vast dust bowl putting thousands of farmers out of work, and many people woke up every morning wondering what, if anything, they might find to eat that day.

Into this bleak picture entered Franklin Delano Roosevelt, the former governor of New York and a man of great optimism. Immediately upon taking office he began to use "fireside chat" broadcasts over the radio to tell Americans that things would turn around for them, the economy would pick up again, and the world would right itself. He spoke to the country in such calm and reassuring terms that people began to regain a sense of hope. Then, before any major improvements could be made in the nation's economic condition, the Japanese attacked Pearl Harbor and the United States found itself pulled into World War II—a war that had already enveloped Europe.

Once again, President Roosevelt calmed the anxiety of Americans when in a nationally broadcast address he said, "All we have to fear is fear itself." Roosevelt used the same calm optimism to face the adversity of World War II that he had used to face the adversity of the Great Depression. But what is truly amazing about this story is that not only did he hold the country together under extremely adverse conditions, but he did this while suffering from an acute case of polio. President Roosevelt could not walk—a fact he hid so well that many Americans weren't even aware of it until after his death.

In spite of having to struggle daily against the increasingly debilitating and painful effects of a crippling disease, Franklin Delano Roosevelt remained calm and optimistic in the face of tremendous adversity. He could have simplified his life and eased his

SUCCESS TIP

When you face adversity in your career, remember the example of Franklin Delano Roosevelt. Stay positive and optimistic, avoid shortcuts and solutions that appear to offer an easy way out, and ask yourself, "What can I do to overcome the adversity I'm facing?" Once you know the answer to this question, go forward. Face the adversity in your life, and do what is necessary to turn the lemons you encounter into lemonade.

SUCCESS TIP

> If success were easy, everyone would be successful. But, alas, success is seldom easy—even for those who are able to make it look easy. There will be bumps, potholes, and detours on your road to success. The best advice I can give to rising professionals facing hard times is this: *Just keep going.*

daily pain greatly by agreeing to use a wheelchair. But the attitudes of Americans toward people in wheelchairs were less enlightened in those days. Roosevelt knew that a president in a wheelchair might be viewed as being weak by America's enemies and even by many Americans (this was in the 1940s).

Consequently, not only did he fight courageously every day to win the war and beat the Great Depression, but he did it while supporting himself with heavy steel braces that bit painfully into his frail and paralyzed legs. The leg braces and the façade of healthful vigor he felt it necessary to maintain only added to the enormous burden of adversity this courageous man faced every day from the day he was elected president of the United States until the day he died at his "Little White House" in Warm Springs, Georgia.

You might wonder why a president from Hyde Park, New York, would have a vacation retreat in the tiny, rural Warm Springs, Georgia. The answer reveals even more about the incredible courage and perseverance of this unique individual. Using his own money, Roosevelt had founded a treatment center at Warm Springs for polio victims. The naturally occurring warm springs that bubbled up in this tiny, out-of-the-way Georgia town had a powerful healing effect on polio victims—especially children. Although serving as president of the United States during some of the country's darkest hours robbed him of the time it would have taken to consistently undergo the physical therapy offered at his own treatment center, Roosevelt wanted to make sure that future generations of polio victims had a chance to receive the treatments. Thanks to Roosevelt, they did.

When you face adversity in your career, remember the example of Franklin Delano Roosevelt. Stay positive and optimistic, avoid shortcuts and solutions that appear to offer an easy way out, and ask yourself, "What can I do to overcome the adversity I'm facing?" Once you know the answer to this question, go forward. Face the adversity in your life, and do what is necessary to turn the lemons you encounter into lemonade.

WHEN TIMES ARE HARD—JUST KEEP GOING

If success were easy, everyone would be successful. But, alas, success is seldom easy—even for those who are able to make it look easy. There will be bumps, potholes, and detours on your road to success. It is when dealing with these inevitable difficulties that your willingness to persevere will help you rise above the competition and propel you past them. The best advice I can give to rising professionals facing hard times is this: *Just keep going.*

Of course, advice such as this is easier to give than take. How, you might ask, do I just keep going when I'm discouraged and broken in spirit—when I want to just give up and quit? First, understand that taking my advice to just keep going will never be easy. It will take self-discipline and commitment to your long-term goals. But you can do it. You can just keep going. Here are some strategies that have helped successful people who were facing hard times: (1) look down the road beyond the difficulties; (2) let go and move on; and (3) stay positive. These strategies are explained in more depth in the following sections.

Look Down the Road

Let's say you didn't get a promotion you really wanted or perhaps a raise you had been working for and probably deserve. Maybe someone else was given a professional recognition award that should have gone to you. Disappointments such as these are going to happen, and when they do it is easy to give in and become discouraged. One way to overcome your disappointment and discouragement is to look beyond the immediate situation to a higher goal in the future.

Applying this strategy is like taking a long trip. Consider this scenario. Your destination is thousands of miles away and you've gone less than a hundred miles when your car breaks down. At this point you can either give up on making the trip, or you can get busy finding a way to get your car fixed. The problems with your car might mean it will take longer to get where you are going, but don't focus on this problem. Instead, look farther down the road to your eventual destination. If it is a worthy destination, focusing on it will put into perspective the temporary setback you are now facing.

Consider the story of Susan. She had worked hard to rise to the top in her profession and was making excellent progress when, with very little notice, the owners of her company sold out to a larger company. She and her fellow engineers became victims of the buyout. The larger company fired the entire engineering team from Susan's company. This was a major setback for a person trying to move up in her profession. Now, not only was she not at the top in her profession, but she didn't even have a job.

Susan was down, but she wasn't out. She looked on the episode as nothing more than a bump in the road. Susan's goal was to make it to the top in her profession. She could do that working with another company. During the years before the buyout, Susan had gained a well-deserved reputation for being good at what she did. When several

SUCCESS TIP

When things don't go your way, don't cry over spilled milk. Let go and move on. The most successful people refuse to hold themselves back by expecting life to be fair and then becoming discouraged when it isn't.

SUCCESS TIP

When life deals you a bad hand, don't allow yourself to get hung up on the unfairness of it all. The fact that the outcome of a situation is unfair is unlikely to change the outcome. While fuming and fussing about how unfair life is, you are wasting energy that could be put into pursuing your career goals. Let go and move on.

companies that were competitors of her former employer found out she was available, they practically started a bidding war for her services.

Susan was out of work for a month, but the offer she eventually accepted from a former competitor turned out to be an improvement over her previous situation. She was able to negotiate both a raise and a promotion. When others might have become discouraged and rushed to take any job they could get, Susan kept her eye on the long-term target farther down the road and turned adversity into opportunity.

Let Go and Move On

When facing difficulties in their careers, some people undermine their chances of success by continuing to look backwards at the problem rather than looking forward at the solution. When things don't go your way, don't cry over spilled milk. Let go and move on. The most successful people refuse to hold themselves back by expecting life to be fair and then becoming discouraged when it isn't. Life isn't always fair, and no matter how hard good people try to make it so, this enduring fact is not likely to change. I hope you will always be fair and equitable with the people on your team and in your organization. I hope you will do everything in your power to adopt policies, procedures, and practices that will ensure as much fairness and equitability in your organization as it is possible to have. But, at the same time, I hope you will understand that bad things do happen to good people, and that in spite of the best efforts of people who want it to be otherwise, life is not always fair.

When life deals you a bad hand, don't allow yourself to get hung up on the unfairness of it all. Yes, you probably should have gotten the promotion instead of the person who did. Yes, you probably did deserve the raise you didn't get. And yes, in both cases, it was unfair that you came out on the wrong end of the situation. But the fact that the outcome of a situation is unfair is unlikely to change the outcome. While fuming and fussing about how unfair life is, you are wasting energy that could be put into pursuing your career goals. Let go and move on.

Stay Positive

The easiest thing in the world to do is fall into despair—become discouraged, dispirited, and negative when things aren't going your way. In fact, the number of people who can stay positive when their world has turned negative is very small. This is one of the reasons learning to face adversity without falling into despair is so important to you in building a winning career. Being able to do what needs to be done when others can't is one of the unalterable keys to success. It's easy to be negative. Any loser can do that, but staying positive requires effort, self-discipline, and courage—characteristics of winners.

So how can you stay positive in the face of adversity? The following strategies can help:

■ *Avoid negative people.*   Some people are just pessimists. No matter the issue they can find negativity in it. Be assertive in avoiding such people. It is difficult enough to stay positive in the face of adversity without having matters complicated further by gloomy, perpetually negative people.

■ *Interact with positive people.*   Some people are like Labrador retrievers—their tails are always wagging whether they should be or not. Such people are blindly and artificially positive. These are not the types of "positive" people you should interact with. The kind of positive people you should interact with regularly and frequently are the kind who face adversity head-on and say, "All right, we have a problem here. Let's see what we can do to solve it." Such people maintain a consistently upbeat, can-do attitude when times are good and when they are bad. There is nothing artificial or fake about their optimism. It is real and founded on substance. Just being around such people will pull you up when adversity gets you down.

■ *Use driving time to regain a positive perspective.*   When facing adversity, the time you spend driving to work, meetings, home, and errands can be invaluable for regrouping; especially when driving by yourself. By regrouping I mean working past the shock or disappointment of adversity, putting things in perspective, letting your positive attitude reassert itself, and beginning to focus on solutions rather than problems.

■ *Use good books, tapes, and CDs.*   Reading good books is an excellent success strategy as is listening to them on tapes or CDs. But when facing adversity, this strategy becomes even more important. "Good books" in the present context refers to books that are uplifting—books that profile others who have faced adversity and overcome it in positive and inspiring ways.

■ *Help someone else who is facing adversity.*   When you face difficult times, one of the best ways to help yourself is to help someone else who is facing adversity. Reaching out to someone else who is hurting can benefit you in several ways including the following: (1) It can show that you are not alone in your distress and, sometimes, that others are facing even worse problems than you; (2) It can take your mind off your problems long enough to give you time to regain a positive perspective; and (3) It can help you transition from a *problem mode* to a *solution mode*.

REVIEW QUESTIONS

1. Explain in your own words the value of perseverance in building a winning career.
2. Explain several strategies for persevering when you feel like giving up.
3. What is the key to facing adversity in your career development?
4. What are the three strategies that will help you keep going when facing hard times?
5. Explain five strategies for staying positive when faced with adversity.

SUCCESS PROFILE

Michael H. Denigan
Engineer

Michael H. Denigan is a mechanical engineer in the Munitions Directorate of the United States Air Force Research Laboratory at Eglin Air Force Base. In this position he applies the engineering skills he first learned in college and developed over time through experience to help give the U.S. Air Force the best munitions in the world. Denigan and his team develop innovative warhead-related solutions for unmanned aerial vehicles. As a result of his team's efforts, when the nation must go to war, certain munitions can be delivered on target without having to risk the lives of pilots and navigators.

Denigan has worked in the development, design, and testing of explosive ordnance for military applications for more than twenty-five years, and in the process built a winning career for himself. Along the way he has applied many of the success strategies explained in this book. His career exemplifies three of these strategies in particular: (1) balancing work and the rest of your life; (2) persevering through adversity; and (3) helping others through mentoring.

Prior to securing his current position with the U.S. government, Denigan worked in a series of increasingly responsible positions for Honeywell, Inc., Gen-Corp Aeroject, and General Dynamics. He worked hard and smart, receiving numerous awards for outstanding performance (e.g., the Honeywell Engineering Week Achievement Award, the Honeywell Stock Recognition Award, the Honeywell Special Achievement Award, the Aerojet Special Achievement Award, and the AFRL/MN Special Act Award). But Denigan had to do more than just work hard and smart in order to succeed. He often had to work long. When it was necessary to work long hours, Denigan did it. But in spite of the long hours, he managed to maintain an appropriate balance between work and home.

As a father of two children, Denigan's obligations are not restricted to just work. He has had to work equally hard and equally smart to make sure that he carves out sufficient time for his wife, son, and daughter. In fact, although he worked for several different employers as he climbed up the career ladder, his family's needs were always a critical factor when deciding whether to accept or decline job offers. There have been times when Denigan could have made substantial career advancements by accepting a position with a new employer, but this usually meant he would have to relocate his family; something he was willing to do only if the move would be good not just for his career, but also for his family.

This was the situation he faced when he accepted his current position with the U.S. government. When a corporate restructuring meant that in order to keep his job with a private sector company, Denigan would have to relocate his family to another

(*continued*)

(*continued*)

state, he faced a major dilemma. His family was well settled in a nice home in a safe community with good friends nearby. Denigan felt that his family's community, friends, and church nurtured and supported them. He determined that relocating his family was not an option. Consequently, he took a courageous step. Without the offer of a job in his local community, Denigan declined jobs in other states.

The responsibility of providing for a family can be scary enough in even the best of circumstances, and, naturally, bearing this responsibility without a job was a daunting prospect. Giving up a well-paying job was like stepping off of a cliff in the dark. You don't know how far you will fall or how hard you will land. However, determined to secure a position in his current community so that his family would not have to relocate, Denigan persevered. After a month of enduring the often harsh realities of the job-seeking process, he finally secured his current position at Eglin Air Force Base.

Having worked as an engineer in the private sector for almost twenty-five years, working as a government engineer required adjustments. However, Denigan persevered in learning how to apply the engineering skills he had honed in the private sector to his new job in the public sector. As a result, he now holds a responsible position in which he uses his twenty-five years of experience to mentor less experienced engineers and to continually improve the quality of the processes he and they work with on a daily basis.

Michael Denigan is an excellent role model for newly hired engineers and for college students who hope to become engineers. Learn from his example how to balance work obligations and family responsibilities, how to persevere through adversity, and how to help yourself by helping others through mentoring.

DISCUSSION QUESTIONS

1. Discuss how you might go about persevering in the following situation: You have been working hard, long, and smart for a promotion in your company, but for the second time, someone else got the promotion instead of you. You feel like quitting.

2. Discuss how you might go about persevering in the face of the following situation: You are a single parent with a good job. Things seem to be going your way. Then suddenly your life is turned upside down by a hurricane that destroys your house and your car. You still have your job, but not much else.

3. Discuss how you might go about staying positive in the following situation: The company for whom you work for has just been purchased by a large conglomerate that does not have the same "family orientation" as the previous owner. Your car is having mechanical problems, and you have just had a major falling-out with your best friend.

ENDNOTE

1. Nancy Lopez as quoted in *Lessons for Success*, ed. Lorraine A. Darconte (New York: MFL Books, 2001), 103.

Appendices

APPENDIX A: RESUMES

A good resume is a key element in landing a great job in any profession. While it is not the focus of this book to detail all of the elements of a well-constructed resume, there are a number of books on the subject available. The reader is encouraged to use such a book as a reference before setting out on an employment search.

Helpful titles from this publisher include the following.

Brown, Lola. *Resume Writing Made Easy: A Practical Guide to Resume Preparation and Job Search.* 8th ed. Upper Saddle River, NJ: Prentice Hall, 2007.

Hanna, Sharon L. *Career By Design: Communicating Your Way to Success.* 3rd ed. Upper Saddle River, NJ: Prentice Hall, 2005.

Harris-Tuck, Liz, Annette Price, and Marilee Robertson. *Career Patterns: A Kaleidoscope of Possibilities.* 2nd ed. Upper Saddle River, NJ: Prentice Hall, 2004.

Robbins, Carolyn R. *The Job Searcher's Handbook.* 3rd ed. Upper Saddle River, NJ: Prentice Hall, 2006.

Sukiennik, Diane, William Bendat, and Lisa Raufman. *The Career Fitness Program: Exercising Your Options.* 8th ed. Upper Saddle River, NJ: Prentice Hall, 2007.

Appendix B: Job-Seeking Tips

1. *Be open to relocating if necessary.* Sometimes it is necessary to relocate to another community in order to find the job you want.
2. *Be positive, proactive, and assertive in your job search.* The job-seeking process can be frustrating because it sometimes involves rejection. Stay positive and keep trying. Be proactive and assertive and you will find the job you want.
3. *Don't be afraid to start at the bottom in your profession and work your way up.* You might have to accept an entry-level position in your profession until you prove your worth. If so, take the job and get busy showing what you can do. Advancement will come.
4. *Don't become frustrated.* It might take numerous interviews before you finally land the job you want. This is normal. Don't let it frustrate you. Stay positive and keep trying until the employer you want to work for says "yes."
5. *Use all resources available to you.* When seeking that first job after college, it is important to use all of the resources available to you. These include the Internet, your institution's career center, friends, and family members.

Appendix C: Interviewing Tips

1. Arrive for the interview ten minutes early and properly dressed.
2. Shake hands firmly and look those who will interview you in the eyes. Introduce yourself and make a point to remember their names when they introduce themselves.
3. During the interview, effect a positive, confident attitude and answer all questions openly and honestly.
4. During the interview, emphasize your positive characteristics and let the employer know that you are a fast learner.
5. If asked to fill out an application, write neatly. Your writing will be a reflection of you.
6. Before leaving an inconclusive interview, ask the employer when a decision about the position will be made.
7. Don't be pushy, but make sure the employer knows you want the job and that if given the opportunity you will do a good job.
8. After the interview, find a quiet place and analyze your performance. If you made mistakes, make note of them and correct them in your next interview. If there were surprises, make note of them and be ready next time.
9. If you are turned down, ask the employer for suggestions that will help you do better next time.

Index